T0275986

Astronomers' Universe

David Seargent

Weird Astronomical Theories of the Solar System and Beyond

 Springer

David Seargent
The Entrance, NSW, Australia

ISSN 1614-659X   ISSN 2197-6651   (electronic)
Astronomers' Universe
ISBN 978-3-319-25293-3   ISBN 978-3-319-25295-7   (eBook)
DOI 10.1007/978-3-319-25295-7

Library of Congress Control Number: 2015957812

Springer Cham Heidelberg New York Dordrecht London

Springer International Publishing AG Switzerland is part of Springer Science+Business
Media (www.springer.com)

For Meg

Preface

As the title of my previous book *Weird Universe* demonstrates, our cosmic home is a strange place. The human mind, accustomed as it is to understanding our familiar surroundings, sets out on an adventure every time it tries to comprehend the broader picture, the wonderful wider universe that is our ultimate physical environment. Here common sense goes out the window! Ideas which seem strange—which *are* strange—are often the only ones that in the end make sense of what our observations and experiments reveal. As Professor Max Tegmark wisely counseled, we should not dismiss theories just because they seem weird to us, lest we dismiss something that would prove to be a real breakthrough in our understanding of nature. Tegmark was speaking specifically about the elusive Theory of Everything when he made this remark, but his statement remains true for lesser theories as well and should be remembered whenever astronomical and cosmological speculations start to look more like science fiction than what we might normally think of as sober fact.

Nevertheless, there is another side to this as well. Just because a theory is strange does not *necessarily* mean that it is on the right track. To assume this would be to go too far in the direction away from common sense. First of all, there are the truly "crackpot" ideas which diverge so far from the overall corpus of scientific discoveries as to be ruled out immediately. What person having even a rudimentary degree of scientific literacy could accept, for example, the "cosmology" of Cyrus Teed who taught that the Earth is hollow and that we live on the inside? Yet, other ideas cannot so readily be dismissed and it is not always easy to know where to draw the line between genuinely crazy theories and those which only superficially appear so because of their counterintuitive nature. This was summed up by the scientist who wondered if physicists living a century from now will look back on some of the leading ideas in contemporary physics and be impressed by the insight of today's scientists or whether they will instead ask what these folk were smoking in the early twenty-first century! Which contemporary theories seem weird because of their deep and counterintuitive insights, and which are truly outlandish?

In this book, we journey through several hypotheses which, for one reason or another, seem strange, out of the mainstream or counterintuitive. Some of these have already been proven incorrect through the accumulation of observational evidence acquired since they were initially put forward. Others remain controversial while still others are widely accepted by mainstream science even though the jury is still out concerning their validity.

Truly crackpot ideas are not, however, included here. All of the hypotheses discussed in the following chapters were at one time put forward as serious explanations for certain astronomical observations by people with credible scientific qualifications. In the majority of instances, the originators of these theories were leading scientists and experts in their field of study. As such, their ideas are not to be lightly dismissed. Even those hypotheses which have subsequently been demonstrated as being incorrect remain valuable. They not infrequently contain an element of truth which may not otherwise have been considered and, additionally, they forced others in the field to take notice of new ideas and approaches which might have been overlooked had no such challenge been presented. The more radical ideas of Fred Hoyle are acknowledged to have exercised precisely this effect—the challenge to "prove Fred wrong" provided the stimulus for quite a deal of research. Even if they serve no other purpose, theories from outside the mainstream at least force us to keep our minds open.

Cowra, NSW, Australia   David A.J. Seargent

The point is that if we dismiss seemingly weird theories out of hand, we risk dismissing the correct theory ... (Professor Max Tegmark).

Acknowledgments

My thanks are extended to the staff of Springer Publishing, especially to Ms. Maury Solomon for encouraging me to consider a book on the "weirder" astronomical theories and to Ms. Nora Rawn for her encouragement and guidance in bringing it to publication. I would also like to offer special thanks to Mr. Michael Mattiazo for his permission to use his image of Comet Finlay. My thanks are likewise extended to Professor Hans Pflug for sending me copies of his transmission electron microscope images of organic structures found within the Murchison meteorite and for providing me with information to which I would not otherwise have had access. My contact with Professor Pflug took place more years ago than I would like to admit, at a time when I was researching the circumstances surrounding the fall of this meteorite. Unfortunately, I no longer know of his whereabouts, but I am most grateful for his assistance back then and for the opportunity to use some of the provided material in this present work. I would also like to thank my wife Meg for her continuing support and for all who have contributed in any way to the preparation of this book.

Last, but by no means least, I would like to acknowledge the courage of all those throughout the years who have gone against the prevailing tide of opinion and put forward "weird" ideas as to the nature of the universe or of some of its constituent objects. Whether these ideas have turned out to be correct (as some have) or whether they have eventually been disproven, they have all served to stimulate further thought and in that way at the very least, have made genuine contributions to our understanding of nature. For that, we can all be thankful.

Contents

1 Is There a Cosmic Web of Life?

If the advance in astronomical knowledge acquired during the course of recent decades has taught us anything at all, it is surely the undeniable fact that we are an integral part of the universe. We are not isolated from the stars and the galaxies. We might even grudgingly admit that the astrologers got it a little bit right after all, though not in the manner that they would have us believe. The position of the constellations and the planets at our birth might not determine whether we will be happy and outgoing or moody and a general pain to work with, but the nature of the universe at large played a very large part in you and I being here at all. The picture that has emerged from relatively recent astronomical research is one in which the entire cosmic environment has played and continues to play a vital and indeed determining role in the existence of life here on Earth. The nature of our home planet, our home star, our location in a relatively quiescent region of our home galaxy and even the immediate cosmic environment of this galaxy itself (viz. its location in a small galaxy group rather than a large galaxy cluster where collisions between major systems tend to strip these of the interstellar material so important to maintain a healthy rate of star formation and to ensure plenty of material for the accumulation of planetary companions of new stars) all conspire to secure our home in the universe.

But what of life itself? Does the universe connect with life on Earth in ways beyond "just" determining the suitability of its terrestrial home? Could life itself have a cosmic connection? Some daring scientific thinkers theorize that indeed it could!

© Springer International Publishing Switzerland 2016
D. Seargent, *Weird Astronomical Theories of the Solar System and Beyond*, Astronomers' Universe, DOI 10.1007/978-3-319-25295-7_1

The Panspermia Hypothesis

Early last century, in 1903 to be more exact, Swedish physicist and founder of physical chemistry, Svante Arrhenius (1859–1927) put forward the radical hypothesis that life is indeed cosmic. He theorized that the seeds of life are carried through space, taking root on any suitable planet upon which they might land. Like grass seed carried in a prairie wind, the germs of life distribute through the universe. Actually, Arrhenius was not really the first to come up with this proposal. As long ago as the fifth century BC, the Greek philosopher Anaxagoras mentioned the idea in his writings, as did the more recent thinkers J.J. Berzelius (in 1834), H.B. Richter (1865) and H. Von Helmholtz in 1879. However, it was through the developed formulation of Arrhenius that the idea truly blossomed into a scientific hypothesis, known as *panspermia* (Fig. 1.1).

This blossoming came forth in Arrhenius' article *The Distribution of Life in Space*, published in 1903, in which he argued that microscopic organisms are theoretically capable of being transported through space by the pressure of stellar radia-

FIG. 1.1 Svante Arrhenius circa 1910 (*Courtesy*: German Wikipedia)

tion. The existence of radiation pressure and its ability to transport very small particles is well established. The anti-solar dust tails of comets and the clearing of star-forming nebula in the near vicinity of bright young stars are proof enough of that, so the basis of the hypothesis rests squarely on good science. Arrhenius argued that particles smaller than 1.5 μm in diameter would be susceptible to this pressure and could be accelerated to high speeds away from the Sun or similar stars. Larger particles are less affected however, so the process could only work for the smallest biological entities. Nevertheless, as bacterial spore fall within the acceptable size limit, and as these can be wafted high into the atmosphere of Earth, it seems reasonable to expect that a certain percentage of these spore could be blown away from the upper atmosphere by the pressure of sunlight and accelerated through the surrounding void of outer space.

Of course, what applies to Earth presumably applies to any life bearing planet. In effect, a planet rich in bacterial life could be thought of as possessing what we might call a "bacterial tail": a plume of spore sweeping away from the planet in a direction opposite to that of the central star. Any planet orbiting outside of the biologically active one would (other things, such as orbital inclination, being equal) periodically pass through this tail, at each passage sweeping up some of the spore which would then filter slowly down through its atmosphere, eventually settling on the planetary surface. Assuming that conditions on that surface were not too hostile, some of these spores might survive and multiply, eventually resulting in a flowering of life on that planet. Eventually, presumably after the passage of many millions of years, the seeded planet would have developed such a teeming biosphere that it would be shedding its own microorganism spore into space—maybe seeding another world beyond its orbit. In this way, as science writer Poul Anderson long ago remarked, a single original life-bearing planet could theoretically seed an entire galaxy. All that is needed is plenty of time and the ability of a percentage of dormant microbial spores to survive the rigors of space for eons of time while remaining capable of revival upon reaching a friendly environment.

Time is something that the universe has in abundance, but the ability of spore to remain viable over the required periods of time, especially considering the constant exposure to cosmic radiation and the other hazards of space, is certainly questionable. The reaction of most scientists was one of strong doubt that bacterial spore could make the journey from one planet to the next without suffering fatal damage to their DNA. So, while it is probably quite widely agreed that spore are indeed wafted into outer space from the upper atmospheres of Earth and other biologically active planets, the general consensus of opinion has traditionally been skeptical that any of these organisms could survive long enough in the space environment to allow this form of panspermia (now known as *radiopanspermia*) to work.

Some critics also raised the objection that the hypothesis does not account for the original genesis of life. Actually, given the belief in eternal matter and the sort of steady state universe prevalent in Arrhenius' day, it might then have seemed legitimate to suggest that life never had a beginning. Like the universe itself, it has always been here! Such an escape route is not available nowadays, in view of Big Bang, cosmic inflation and such like.

Also, for the hypothesis to be capable of accounting for life on Earth, it must assume a biologically active Venus. If Earth picked up spores on their journey away from the Sun, they could only have come from Venus or Mercury (or from comets and Sun-approaching asteroids, but that hypothesis is a later addition to the original). Mercury does not appear a likely source and, in any case, transportation from there would involve a longer trip and a consequent multiplication of the dangers encountered along the way.

Another difficulty is raised by the fact that radiopanspermia relies on the repulsive force of stellar radiation. This means that it is not a good way of accounting for life on any Earthlike inner planet of any Solar System. Planets such as Earth—traditionally considered to be the most likely places where life might be found (not surprisingly—after all, we are here!)—exist in regions where the pressure of stellar radiation is quite strong. Therefore, one would expect bacterial spore to be blown away from those planets deemed to be the potentially most life-friendly.

Radiation pressure is not, however, the only way that we can imagine life to be distributed through space. Another and somewhat more promising means of cosmological transportation makes use of the inside of boulders several meters in diameter. This form of the hypothesis is known as *lithopanspermia* and it solves in one fell swoop several of the difficulties faced by the earlier model.

For a start, rocks do not depend on the acceleration acquired from stellar radiation pressure. A rock can wander through space along a wide variety of orbits. It can venture close to a star; even falling into it if the periastron of its orbit has been reduced to a distance from the center of gravity of its system smaller than the radius of that system's central star. Not a happy outcome for the rock, but at least it could accomplish something that naked bacterial spore driven outward by the pressure of radiation could not.

A less extreme accomplishment would be for the rock to collide with one of the inner planets. If that planet possessed an atmosphere worthy of the name and if the rock was large enough, strong enough and hit the planet's atmosphere with a sufficiently low velocity, some fragments of its inner parts could survive to the surface. Of course, this is happening all the time on Earth. We call those fragments meteorites. Moreover, we know from experience that, although the flight through Earth's atmosphere raises the temperature of the surface of a space rock to incandescence, the flight itself is too brief for the heat to penetrate very far beneath the surface and therefore the interior of a rock of sufficient size remains cold throughout the entire atmospheric trajectory. Contrary to what is popularly thought, meteorite fragments are not "red-hot and glowing" when they fall. Those with a high metal content might, for a short while, be too hot to hold but the more common stony kind have temperatures ranging from pleasantly warm to freezing cold. After all, it is not unheard of for a meteorite, even one falling to Earth on a hot day, to be coated with a layer of frost.

If certain meteorites really do harbor bacterial spore, these will be shielded from cosmic radiation by the meteorite's rocky body in a way that naked spore open to the rigors of space will not be. A few meters of rock lying between the spore and outer space can do wonders for the former's survivability.

While we have been talking about "spore", it is not entirely beyond the bounds of possibility that actual functioning organisms could be transported between worlds in this way. Although microorganisms requiring air and/or sunlight are ruled out, something resembling the very slow-metabolic *bacillus infernus*, an organism which flourishes deep within the crust of our planet, might well remain active during a long trip deep within a space rock.

Some More Exotic Versions of the Hypothesis

The very idea of panspermia might seem pretty weird to most folk, but some versions of the idea are really out of the left field.

Perhaps spore is being transported through space within something more sophisticated than a lump of rock. The possibility of bacterial spore remaining viable is something taken seriously by those designing our space probes and planning missions to other planets. We certainly don't want to contaminate other worlds with Earth bugs. But what if ancient Earth had long ago been visited by an alien probe from a civilization that was not so careful? What if this probe landed on Earth about four billion years ago? Indeed, what if an occupied spaceship set down on this planet soon after its formation and accidentally left some bacterial spore behind? Could that act of carelessness have been responsible for life on our planet?

Although this hypothesis is highly improbable and is, in any case, almost impossible to verify or falsify, it was put forward as a serious suggestion by the well-known astronomer and cosmologist Thomas Gold back in 1960. Gold probably did not believe it, but he apparently considered it worthy of mention as a possibility.

An even more daring suggestion was made by none other than Francis Crick, co-discoverer of the double helix, for which he won the 1962 Nobel Prize for Physiology or Medicine together with his colleagues James Watson and Maurice Wilkins. Together with Leslie Orgel, Crick proposed that life on Earth may have been purposively planted here by an advanced extraterrestrial civilization. One version of this *directed panspermia* hypothesis proposes that an advanced civilization might direct capsules containing life seeds toward regions where new stars are forming. By the time these capsules reach their destination, planets should have formed around many of these young stars and at least a few of these might act as fertile fields in which the microbial passengers carried by the capsules could take root.

While this hypothesis might make a good theme for a science fiction novel, in reality it is seriously lacking in evidence. Indeed, one might suppose that if the sort of super civilization hypothesized here existed within our galaxy, there should be some other evidence of its presence. Assuming that the civilization that seeded Earth continues to exist, may we not see evidence of galactic engineering, or structures within the galaxy that defy explanation in purely natural terms? Pulsars were once thought to be beacons constructed by a highly advanced civilization and designed to act in the manner of interstellar lighthouses before these became known to have formed purely by physical processes. Of course, one could argue that the hypothetical civilization that seeded our world became extinct sometime after this event. Maybe its desire to send the seeds of life into space was a sign that it was already dying and that this project was a way of perpetuating a biological future for the Galaxy. One would like to think, however, that a civilization so advanced could find some other way of perpetuating its existence.

Be that as it may, this hypothesis strikes a difficulty in the form of galactic evolution. Galaxies are not static systems. Generations of stars are born and die within them and each generation leaves its special legacy in the form of increasing amounts of heavy elements (traditionally but rather inaccurately termed "metals" by astronomers) within the interstellar medium. As it is from this medium that new generations of stars are formed, it is inevitable that each succeeding generation of stars contains an increasing proportion of these heavier elements; ashes, so to speak, of their predecessors. Although not constant across a galaxy, the general evolutionary trend is for galaxies to increase in metallicity over cosmic time. Because living organisms require a relatively high concentration of heavy elements in their environment, a galaxy does not become life-friendly until it reaches a certain state of development. Our home galaxy obviously reached that stage about 4.6 billion years ago when the Sun was formed. Or at least, based on the very existence of life on Earth, the part of the Galaxy in which the Sun formed had then reached that stage of metal enrichment. However, the Sun appears to be a little more enriched with metals than most stars of its age and type, so there is reason to think that its galactic nursery was somewhat ahead of the average in its holdings of life-friendly elements. In short, the Sun (and with it of course, the Earth) was an early starter in the race toward life friendliness. That does not preclude the possibility of life-bearing planets older than Earth or, for that matter, of civilizations older than ours. We are not constrained to believe that our Solar System was the *very* first to possess a life-friendly metallicity. But it does cast doubt upon the existence of a civilization of such great age as to be capable of bio-engineering the Galaxy at the time the Sun was just forming.

The main difficulties with the directed panspermia hypothesis, and indeed with any form of panspermia are the twin problems of lack of evidence for the existence of advanced life (or, for that matter, any life) beyond Earth and the fact that the hypothesis does not provide an explanation as to how life started in the first place. Even if the theory proved to be correct, it does not tell us how life first appeared in the universe. As mentioned earlier, only if life is eternal in a steady state universe could panspermia be considered complete in any sense. But that possibility is, as already noted, precluded by those cosmologies supported by the weight of contemporary evidence.

"Soft" Panspermia

Whether in its moderate or more adventurous forms, panspermia in the sense of the word which we have been using here, does not win many adherents amongst scientists. The possibility of dormant organisms being wafted from one planet to the next or being carried within meteorites is not rejected, but it is fair to say that the majority of scientists working on the issues surrounding biogenesis do not see this as a major process. If it does really happen, it is a secondary rather than a primary consideration in the opinion of most workers in the field. The principal exception to this line of thinking is Chandra Wickramasinghe and, previously, Fred Hoyle, about whom more will be said later in this chapter.

For the present however, let's look at a very modified version of the hypothesis sometimes called *pseudo panspermia* or, alternatively, soft panspermia. In contrast with the varying versions of hard panspermia considered thus far, this hypothesis does not propose that life itself was transported from elsewhere to Earth, or to any other life-bearing world that may exist. Life per se is indigenous to the planet on which it flourishes. What is transported from elsewhere is not the organisms themselves but the chemical compounds of which these organisms are composed. Proto-life chemistry, not life itself, is brought from somewhere else according to the soft form of the panspermia hypothesis.

While this hypothesis does not explain the origin of life, it does show why the necessary chemicals existed on the early Earth. The existence of so-called organic compounds on the early Earth became a problem as continued examination of the most ancient rocks failed to reveal the sort of reducing atmosphere that had previously been thought to surround our planet in its infancy. If, as widely thought until the latter years of last century, the atmosphere of our planet circa four billion years ago consisted of such gases as methane and ammonia mixed together with hydrogen, the presence of complex organic molecules presented no problem. In 1953, the famous Urey-Miller experiment, in which electric sparks were fired through a mixture of methane, ammonia and hydrogen within a flask containing water, clearly demonstrated that organic molecules were readily synthesized in this environment. If the gas mixture in their experiment matched that of the early terrestrial atmosphere, lightning and ultraviolet radiation from the Sun must surely have synthesized a great deal of organic material. Washed down into the ancient lakes and oceans, this organic soup seemed the perfect place for life to find its toehold.

The only problem with this promising hypothesis was the complete failure to find any evidence that the most ancient rocks were ever exposed to such an atmosphere. On the contrary, all evidence pointed in the opposite direction. The early atmosphere was at most neutral; possibly weakly oxidizing. Under conditions such as these, organic compounds simply could not form, no matter how much ultraviolet radiation streamed from the Sun and how much lightning flashed through the skies. Following this line of reasoning implies that Earth should be devoid of organic compounds; a striking contradiction to the facts on any assessment.

Perhaps therefore this planet's stock of organic compounds—the chemical precursors of living organisms—did arrive here from outer space. Superficially, this sounds a tad farfetched, but evidence in its favor has been steadily accumulating over the years. Much of this evidence has come in the form of organic substances found in meteorites, especially (though not exclusively) in members of that class of stony meteorites known as carbonaceous chondrites. Some surprisingly complex, not to say biologically significant, compounds have been extracted from meteorites of this class, but a perpetual difficulty that has haunted this research is the matter of distinguishing between those compounds that are indigenous to the meteorite and terrestrial material which has contaminated the stone after it reached the ground or even managed to get drawn into the object as it plummeted through the atmosphere. One problem is that the biologically most interesting meteorites are quite porous. In space, these pores are, of course, vacuous but once Earth's atmosphere is entered, air is drawn into them and anything which might be floating around in the air gets dragged in as well.

Fortunately, terrestrial organic material shows a specific ratio of carbon isotopes and in 2008, analysis of organic compounds extracted from fragments of the Murchison meteorite—a carbonaceous chondrite that fell in Australia in 1969—revealed a different and entirely non-terrestrial isotope ratio. This is strong evidence that these compounds at least are not contaminants. And they make an interesting group: amongst the compounds identified in this meteorite are the biologically relevant molecules uracil and xanthine. The first of these is an RNA nucleobase (Fig. 1.2).

Equally as remarkable was the identification, announced the following year, of the amino acid glycine in dust particles collected from the coma of comet 81P/Wild 2 in January 2004 and subsequently returned to Earth.

More recent discoveries include the discovery of complex organic matter in cosmic dust and, in 2012, the detection of the sugar molecule glycolaldehyde in the infrared source *IRAS 16293-2422*, a protostellar binary some 400 light years away. This molecule is required for the synthesis of RNA. Then, in 2013, the Atacama Large Millimeter Array confirmed the presence of a pair of prebiotic compounds, cyanomethanimine and ethanamine in a molecular cloud 25,000 light years from Earth. The first of these is thought to be a precursor to adenine, one of the four nucleobases forming the rungs in the ladder-like structure of the DNA molecule, while the second is believed to be of importance in the formation of the amino acid alanine. It now seems that perhaps one fifth of the universe's stock of carbon is in the form of rather complex organic materials, many of which are of great biological interest.

Fɪɢ. **1.2** A fragment of the Murchison Meteorite (*Courtesy*: Museum of Natural History, Washington)

Many of these organic molecules form on the surfaces of cosmic dust particles; the same types of particles which snowball together to form the boulder-sized bodies which constitute an early step in the process of planet formation. Perhaps the "sticky" coating of organic molecules plays an important role in causing these particles to stick together and in this way directly aids in the process of planet building. Furthermore, although the heat and compression of a newly formed planet must destroy such complex molecular structures, there is no reason to think that smaller planetesimals will not preserve these compounds. In fact, the conditions inside a planetesimal large enough to generate sufficient heat to melt ice but too cool to destroy organic substances may give rise to some very interesting organic chemistry indeed, but more about this in a little while. For the present, let us just consider how the influx into the Earth's early atmosphere of dust coated with organic substances and the sporadic arrival of organic-rich meteorites such as carbonaceous chondrites could have enriched our planet with loads of material from which the chemistry of life arose. Although it may have once appeared to be such, this scenario is no longer seen as being a weird hypothesis. We may or may not wish to include it under the umbrella of panspermia, but it certainly does not look eccentric in the light of present knowledge.

Slightly Harder Panspermia

Analysis by David Deamer and colleagues of some of the organic molecules retrieved from the Murchison meteorite revealed one particular specimen that held great interest, not merely because of its complexity but also because of its behavior when introduced to water. The molecule consisted of a long chain whose individual links could be regarded as simpler organic molecules having differing characteristics. Some of these are hydrophilic (literally water loving) while others are hydrophobic (literally water fearing or water hating). When this molecule is brought into contact with liquid water, its hydrophilic segments are attracted to the water molecules while its hydrophobic segments are repelled by them. The result is that the long chain molecule rolls itself into a ball having the hydrophilic segments on the outside, in contact with the water, and the hydrophobic segments on the inside sheltering, so to speak, from the water behind the protective wall of the former. This arrangement forms a bilaminar barrier between the inside of the globule and the surrounding water. Bilaminar globules tend to permit molecules to pass through into their centers where they can concentrate and react further with one other, building into larger molecules of even greater complexity. To this degree, they act a lot like very, very, simple biological cells. The intriguing question is whether these are in some way the precursors to true biological cells. Is it possible that, by concentrating molecules into the centers of these globules, eventually self-replicating nucleic acids emerged within them, transforming them from being mere globules into genuine biological cells, albeit ones of extreme simplicity?

The situation is, no doubt, a lot more complex than this, but there may be evidence that cell-like vesicles were indeed present in the parent bodies of at least some carbonaceous meteorites. Back in the 1980s, Professor H-D Pflug examined fine sections of the carbonaceous meteorites Murchison, Orguil and Allende with a transmission electron microscope and found that a significant percentage of the organic material within these bodies was in the form of tiny structures having a variety of morphologies. These structures were of micron dimensions and smaller and constituted such a large part of the carbonaceous material that contamination could effectively be ruled out. After all, the structures were literally like nothing on Earth; nothing, at least in that size range although some of them bore superficial resemblance to iron-oxidizing bacteria, albeit about an order of magnitude smaller (Fig. 1.3).

Intriguingly, many of the structures appeared to be just like tiny cells and even showed evidence of the sort of bilaminar membranes about which we have been speaking. Just as intriguing is the evidence that these were not static objects; some of them apparently multiplied. This is strongly hinted at by the colonies of cell-like structures and the appearance of what seem to be "buds" on the side of some of the isolated cells. If these features are what they appear to be, it would seem that the cell-like structures absorbed

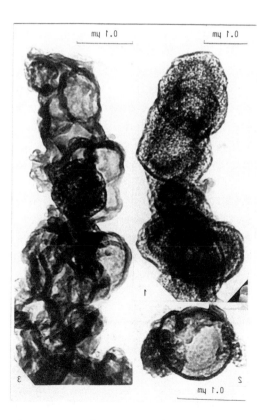

Fig. 1.3 Chain of microvesicles in Murchison Meteorite (Transmission electron microscope image. *Courtesy*: H-D Pflug)

material from their surroundings and reproduced by budding into colonies. Even if they did not possess any genetic coding, this process at least resembles the growth and multiplication of very simple living organisms. Is it possible that some of these cells (Pflug termed them microvesicles, a somewhat less contentious term) might have ended up absorbing some of the biologically interesting compounds mentioned earlier, the ones which now form part of protein or nucleic acid molecules, and brought them into a greater degree of contact than they would have experienced had they simply floated around free in a more or less dilute aqueous solution? This probably did not happen within the body of the meteorite parent itself or else true cells, not merely microvesicles, would presumably have been found within the meteorite fragments. Although some would dispute this statement and insist that genuine cells have indeed been found in meteorites, the more cautious approach favors the process taking place after the meteorites arrived at the surface of Earth or another planet. According to this line of thinking, very early in the life of the Solar System, some of the carbonaceous meteorites may still have been sufficiently fresh to contain viable microvesicles or even a certain amount of liquid within their pores. If these came down in a suitable environment, the vesicles might have continued multiplying and developing in their new home. One might even speculate that the best place for such an active-microvesicle-carrying meteorite fragment to land was in a pond of water already thick with organic molecules supplied by the general rain of such substances from the incessant bombardment by meteorites and cosmic dust experienced by the inner planets during the turbulent infancy of the Solar System. In short, the food for the microvesicles was already there awaiting their arrival; food already delivered by other meteoritic suppliers.

Are We Martians?

Although most meteorites are believed to come from asteroids, some are also known to have originated on the moon and Mars. One originating on the latter world, discovered during an expedition in 2009/2010, proved of special interest. A vein of clay within the meteorite was found, by biologist James Stephenson of the NASA Astrobiology Institute and the University of Hawaii, to contain boron oxide. Now, that may not seem very exciting to most of us, but to researchers into the thorny problem of biogenesis, it was breathtaking. Boron oxide is now thought to have been an essential ingredient in the stabilization of the first genetic material. Most scientists believe that this was not DNA but was instead the simpler RNA, an essential part of which is the ribose molecule. This is a ring-shaped structure and, chemically speaking, is difficult to create from scratch. To form at all, it requires stabilizing substances, one of which is boron oxide. But the problem is, the atmosphere of the very young Earth is thought to have been unsuitable for the formation of this substance. Therefore, if life depended on the existence of RNA and RNA could not exist without boron oxide, how is it that there is life on Earth? It should not have been possible.

This is why Stephenson became excited about his discovery in the Martian meteorite fragment that he was studying. His search of the meteorite was sparked after he came across an article by American chemist, Steven Benner of the Wesheimer Institute of Science and Technology in Florida. This article described the laboratory synthesis of ribose by means of oxidized boron. Realizing that this was something that would hardly have happened naturally on Earth, Stephenson wondered if it might have nevertheless been possible on Mars, hence his interest in the little fragment of that planet found during the recent expedition. If boron oxide could be shown to exist on Mars, it might be argued that RNA first appeared on that planet and was subsequently carried to Earth by Martian meteorites. Furthermore, it is now thought that boron oxide is not the only unearthly chemical required for ribose synthesis. Chemists have also found that molybdenum oxide is also important to the process. This chemical is only possible in very dry environments and is most unlikely to have been around on Earth during, what is believed to have been, the very wet period coinciding with the appearance of the first terrestrial organisms. On the other hand, the far drier Mars is likely to have harbored plenty of it, so it is not unreasonable to suppose that this substance also came to Earth from our planetary neighbor via the delivery vehicles of meteorites.

It would appear that there is a fine line here. Life, or at least the sort of life we know on Earth, requires water, yet the presence of ribose as an apparently essential factor in life's existence equally requires waterless environments and, as such, places a limit on just how wet the birth place of life must have been. A truly watery world with extensive and deep oceans, it would seem, is not the place for life to gain a toehold. What is needed is a principally dry planet with some small lakes and puddles. Such a planet would probably be quite small; too small to retain much water for long periods. Yet, although a world such as this may be ideal as a nursery for the chemical processes preparing the way for life, its small size probably means that it will not be tectonically active for long enough to maintain a permanent magnetic field, needed to shield it against the solar wind and thereby help prevent its atmosphere from being eroded away by the bombardment of energetic solar particles. Furthermore, its gravity would likely be too weak to retain a thick atmosphere for significant periods of cosmic time. If life is to really get underway, this pre-biotic chemistry needs to be transported to precisely the oceanic environment that is so detrimental to its genesis.

We could go a step further along this line of reasoning. It is widely thought that Mars is undersized because the presence of giant Jupiter scattered much of the material from which it would otherwise have incorporated into its formation. The asteroid belt consists of material that, together with much other matter that was flung far and wide, presumably would have gone into the making of a larger Mars, had Jupiter not been present. If it had been larger, Mars may truly have been another Earth, complete with thick atmosphere and widespread oceans. Ironically however, that might not have resulted in two planets populated by thriving biospheres. It might have meant that both Mars and Earth would have been lifeless worlds! At the other extreme however, had Jupiter been significantly larger, Mars might not have been able to form at all. According to this line of thinking, Earth would probably have remained barren. The disturbing influence of an oversized Jupiter would most probably have meant that Earth, if it formed at all, would have possessed dimensions more Mars-like than those we know today. Maybe the prebiotic processes envisioned for Mars would then have occurred on Earth instead. But if Earth was the

"Mars" of that scenario, there would have been no "Earth" to which the chemicals of life, once transported there, could have blossomed into life. Venus would not be a good prospective candidate for sure.

The line of thought followed by Stephenson is interesting, but some have gone even further along this road of speculation and wondered whether the early Mars brought forth something even more complex than the components of RNA. Might primitive life itself have appeared on the Red Planet and is it possible that Martian meteorites brought, not simply the chemicals of life but life itself to its blue neighbor? For this to be possible, microorganisms must have withstood two great shocks; the blast into space of Martian rocks following an asteroid impact on that planet (which is, presumably, how Martian meteorites were launched in the first place) and the atmospheric entry and impact of such rocks on Earth. In addition to these of course, there must also have been an intervening period of indefinite length as the microbes' host rock drifted around in outer space.

The possibility that dormant microbial spore can exist deep within a boulder-sized body in space has already been raised and does appear to be possible for quite extended periods of time. As for the ability to withstand severe shocks, a recent experiment (with Martian meteorite panspermia specifically in view) was carried out by a group of German, Russian and English students with encouraging results. These researchers placed bacterial spores and blue-green algae between layers of rock having similar properties to Martian meteorites and subjected them to pressures similar to those encountered by a meteorite crashing to Earth. Both the spore and the algae survived and continued to grow following their ordeal.

If the Martian connection turns out to be valid, does that imply that the earlier suggestion that the organic material and microvesicles found in carbonaceous meteorites might have played a role in terrestrial life was a dead end? After all, carbonaceous meteorites almost certainly do not hail from Mars.

Not necessarily. Despite its smaller size, Mars is probably hit by asteroidal fragments more frequently than our own world, courtesy of its location relatively close to the main asteroid belt and the greater number of asteroids that cross its orbit. The same is true of short-period comets of the Jupiter family, few of which cross Earth's orbit but a considerable number of which venture into the region between Earth and Mars. Despite the apparently greater suitability of Mars (compared to Earth) for RNA formation, there is no reason to think that it was any more likely to have had an indigenous store of organic materials. Organic materials do not appear to be "native" to the inner Solar System, so both Martian and terrestrial organics were presumably transported to these worlds from the cooler, outer, regions of the planetary system via fragments of carbonaceous asteroids and comets. If a percentage of this organic material arrived on Earth in the form of microvesicles, presumably the same applies to Mars and, if Stephenson is correct, these stood a better chance of absorbing RNA and becoming true biological cells than their counterparts on Earth.

The Stephenson scenario nevertheless may present a problem for the widespread opinion that Jupiter's moon Europa harbors living organisms in its underground ocean. Biologically speaking, Europa appears to have the same problems as the early Earth, i.e. too much water! On the other hand, because Europa lies beyond the orbit of Mars, it might have picked up Martian spore via radio-panspermia. Or, if that seems a bit too farfetched, one could always point to the massive Jupiter attracting Martian meteorites and conjecture that some of these may have struck Europa. The lack of any appreciable atmosphere on this moon might have even been advantageous. Small Martian rock fragments would not have been completely consumed as meteors, as many surely were on Earth. Moreover, the weaker gravity of Europa would not mean such high-velocity crashes as on Earth. We might imagine Martian meteorites landing on the icy surface of Europa and slowly sinking down through the ice into the underlying water. Maybe, if life was transported to Europa in this way, the places where it took root (and maybe where it might be found today) lie within smaller lakes and ponds within the ice, rather than in the deep and extensive ocean believed to exist beneath the icy exterior.

Another issue potentially raised by this scenario follows from the consequence that, if it is true, any Martian life would be essentially the same as terrestrial life. Should life be discovered on Mars and found to be basically the same as the familiar life of our home planet, three explanations for this would immediately present themselves and it would be rather difficult to determine which one was correct. Would such interplanetary similarity imply that life (or at the very least, carbon-based life that uses water as a solvent) is essentially the same wherever it is found, or, that life was transferred from Earth to Mars, probably via the agency of the solar wind, or that life travelled in the other direction a la Stephenson's scenario. Comparison of the organisms alone could not determine the correct answer. Other factors, such as those presented by Stephenson in his argument, would need to be examined and their importance weighed.

In this context, we may note an interesting paper by Nora Noffke of the Department of Ocean, Earth and Atmospheric Sciences at the Old Dominion University in Norfolk. This was published in February 2015 in *Astrobiology* and analyzed features found by the Curiosity rover in sandstone beds associated with the Gillespie Lake Member, a portion of the 5.2 m (about 17 feet) thick Yellowknife Bay rock succession within the Gale crater on Mars. This feature has been interpreted as an ancient playa; a type of lake that temporarily fills with water and is quite common in certain arid and semi-arid regions of Earth. Gillespie must be younger than 3.7 billion years, as determined by the age of the Gale crater, and was found to contain some very interesting structures on a scale of centimeters and meters. According to Noffke's analysis, these look suspiciously like features found in the sedimentary rocks laid down in ancient playa lakes on Earth. The really interesting part is that the terrestrial structures are known to be the work of microbial life. Collectively, the structures are known as microbially induced sedimentary structures or MISS and include such recognized formations variously known as erosional remnants and pockets, mat chips, roll-ups, desiccation cracks and gas domes. If the examples of these features found on Earth have a microbiological origin (as in fact they do!) it would seem reasonable to conclude that the ones on Mars have a similar origin. It might also be seen as a hint that the same general types of microbes responsible for the terrestrial formations were also responsible for the Martian ones.

Even more suggestive, in Noffke's opinion, is the apparent non-random distribution of the MISS-like structures found by Curiosity. From the spatial associations of these features, and their appearance of succession, it seems that they changed over time. On Earth, this type of spatial association and temporal succession would be interpreted as being the record of the growth of a microbial ecosystem that initially thrived in pools only to perish as the latter completely dried up at a later date. The features known as erosional pockets, mat chips and roll-ups are the results of erosion of the ancient microbial mat-covered sedimentary surface by water. Subsequently, channels caused by flowing water cut deeply into the ancient microbial mats, leaving only the erosional remnants behind. Desiccation cracks and gas domes later arise during the final period of atmospheric exposure. Further in-situ observation of the Martian features is, of course, required before any firm conclusion is reached as to whether these really are MISS formations, but if Noffke is correct in this tentative identification, not only will the case for early Martian life be essentially proven, but there will also be a strong hint that this life strongly resembled that which inhabited the early Earth. While that would not prove that life migrated from Mars to Earth, it would at least be consistent with that hypothesis.

Although a side issue in relation to our present discussion, we might also note that the question of a possible Martian origin of terrestrial life may raise moral issues if the prospect of planting a human colony on Mars ever becomes serious. While there are those who would love to see this happen, others are less convinced and some people over the years have been openly hostile to the whole issue of human space travel. C.S. Lewis, for instance, wasted no words in his poem *Prelude to Space*. Admittedly, at the time when Lewis composed this poem, Mars was almost universally believed to be the home of, at the very least, relatively complex plant life and even the notion of intelligent Martians was not entirely confined to the comic books. Lewis may also have had in mind the sort of scenario given fictionally in Olaf Stapleton's *Last and First Men* in which humanity in the far distant future shifted its place of abode to Venus, deliberately wiping out the native race of that world who had been, understandably, not too happy about seeing their planet being taken over by technologically superior humanity. This is, of course, just projecting into a space-faring future the attitude prevalent in a sea-faring past. One could ask a full-blood

Tasmanian aborigine about this ... except that one cannot find any today. They were on the wrong end of British colonialism! We might hope that the human race has morally improved a little since those days, but the fear remains that if humans can commit the sort of crimes, to which History attests, to one another, what might a future colonizing power do to creatures with six legs and four eyes?

Of course, this will remain only hypothetical for the foreseeable future as we can be pretty sure that no other sentient races are within reach of Earth, given the technology of today and, it would seem, many tomorrows. But the moral question as to how (or even whether) we should explore a planet that harbors any sort of life at all may not be hypothetical. If life ever got a toehold on Mars, it is probably still there today, even if only in the form of microbes. Carl Sagan, though a champion of space exploration, was not too far removed from Lewis when it came to life-bearing worlds. He famously stated that although a sterile Mars would be ripe for human colonization, if any form of Martian life existed, then "Mars should be left to the Martians; even if they are only microbes." With deference to political doctrines (the Munro doctrine, Brezhnev doctrine and so forth), let's call this the "Sagan doctrine". According to the Sagan doctrine, we should keep our hands off any planet that harbors life, no matter how primitive that life may be. Part of the reasoning behind this doctrine is in recognition of the great value in preserving, in undisturbed form, any new form of life that we might discover. But what happens to the Sagan doctrine if Martian life is not a new form of life at all? Might it not be argued—against the doctrine—that the colonizing of the Red Planet by emissaries from the Blue one is simply Martian life coming home? It could be the Martians left for Earth as microbes and returned home as humans.

This can be left for the ethicists of the future to worry about. Hopefully, by the time the first spaceships carrying human cargo leave for Mars (if, indeed, that day ever arrives) we will know whether Stephenson was right and we may even have finally settled the question as to whether Mars does indeed have life. How we handle the answer to both these questions may decide the future of our exploration of this fascinating neighboring world.

A Biological Universe?

During the closing decades of last century, the thesis of panspermia took a rather extreme turn which many saw as being tantamount to leaving the road altogether in the theory of Fred Hoyle and Chandra Wickramasinghe. Not their hypothesis that diseases arrive from outer space (though we will be looking at this in due course) but the more general view of these scientists that biology is nature's way of synthesizing complex organic molecules within star-forming and planet-forming regions. These complex molecules, Hoyle and Wickramasinghe argue, are essential if cooling preplanetary material is to reach thermodynamic equilibrium. They argue that at high temperatures, thermodynamic equilibrium favors the existence of as many gas molecules as possible; a situation which leads to the dominance of inorganics such as carbon monoxide and nitrogen. But at significantly lower temperatures, the positive free energy values obtained in the formation of methane and ammonia favor the reduction of the abovementioned inorganic species into ammonia and water. Hoyle and Wickramasinghe however, argue that the required reactions are too inefficient to proceed in a purely inorganic situation. Cooling preplanetary material therefore goes out of thermodynamic equilibrium. It is here, they argue, that biology comes into play. The required species are far more easily generated by biological processes than by inorganic ones. These scientists write that therefore, "Biology is nature's way of moving much closer to thermodynamic equilibrium than would otherwise be possible." (*Living Comets*, p. 76) They add that if this were not so—if thermodynamic equilibrium could be reached without biological processes—nature would have taken this course and life would never have appeared (Figs. 1.4 and 1.5).

Life, on this radical view, is not some secondary epiphenomenon of cosmic evolution. It is involved in the very architecture of the universe.

In answer to such objections as "What about the Urey-Miller experiment?" Hoyle argued that as the methane and ammonia found on Earth are essentially, and in the final analysis, the products of biological processes, should anyone be surprised that an experiment which employed them as its ingredients ended up with molecules that were biologically significant?

FIG. 1.4 Sir Fred Hoyle (*Courtesy*: Donald Clayton)

FIG. 1.5 Professor Chandra Wickramasinghe (*Courtesy*: Wikimedia)

Most astrophysicists would nowadays acknowledge the important role played by complex organic molecules in the wider universe, but what we might term the "orthodox" view is that these build up as a sort of tarry mantel around silicate cores through the action of stellar radiation. The "core/mantel" model of cosmic dust particles is the one most widely accepted by the astrophysicists of today. Complex molecules, having formed within the mantels of dust particles, can then be chipped off into space by exposure to cosmic rays and stellar radiation, becoming part of the surrounding gaseous medium. But this process, if Hoyle and Wickramasinghe are correct, simply does not work as required.

What we might call the orthodox view of the synthesis of organic material involves the presence of a form of the Fischer-Tropsch Process whereby carbon monoxide and hydrogen combine in the presence of a catalyst to form hydrocarbons and water vapor. This is supposed to account for the mantel of organic material that builds up on silicate grains in interstellar dust, with the latter acting as the catalyst. Yet Hoyle has strong doubts that this process is efficient enough to do the trick. The process has been employed industrially in the manufacture of synthetic oil, but the fact that the world continues to rely on natural oil reserves drawn from one of the most politically unstable regions on the planet surely says something about the efficiency of the FTP. If it was truly efficient, why is it not in use in all countries with the required technological resources? Why do we still rely so heavily on extracted oil?

Ironically, it was through the work of these two scientists that the core/mantel model was developed, but while most other astrophysicists accepted it, Hoyle and Wickramasinghe were never wholly convinced that they got it right. The observational tests, such as extinction of starlight at differing wavelengths as it passed through regions of interstellar dust, were close to what the model predicted, but still not exact. At one point, these researchers even proposed hydrogen ice as a major component of interstellar grains, but soon abandoned that possibility because of the extremely low sublimation point of this material.

It was only after many years that Hoyle tried to match the results with dried microorganisms. Remarkably, he found the match to be just about perfect. Indeed, the fit was even better than the one credited as support for the core/mantel model. The fit to interstellar extinction in the Cygnus region, for example, was striking. The conclusion appeared obvious to Hoyle: interstellar dust is composed of the desiccated corpses of microbes.

Moreover, biology also seemed to provide a very good match for the infrared spectrum of warm interstellar dust in the region between about 8 and 13 μm. Hoyle and Wickramasinghe demonstrated this in a very simple and direct way. They first of all went down to the nearest river with a bucket and scooped out some water, after which they cultured the microorganisms found therein until they had a sufficient number to allow determination of their infrared properties. All that was needed then was to compare these properties with those of interstellar dust. The example of the latter chosen by these researchers was the star-forming region in Orion near the four youthful hot stars known as the Trapezium. Once again, the agreement between the curve predicted for microorganisms on the basis of direct observation and that found for the Orion Nebula in the Trapezium region was almost perfect (Fig. 1.6).

Needless to say, this theory was simply too weird for most of the colleagues of Hoyle and Wickramasinghe to accept. One critic stated that the match between dried microorganisms and cosmic dust was a poor fit, despite Hoyle's claim to the contrary. Hoyle did not mince words in his reaction to this critic. "He is a liar, they do match" was his blunt reply.

Hoyle saw those who scorned this theory as being too narrow-minded in their views about life. They were Earth chauvinists, in other words. Nevertheless, his criticism was probably not entirely fair. After all, Earth and similar planets have plenty of ecological niches for life to occupy, but the thought of microbial organisms existing in such number throughout the universe as to constitute

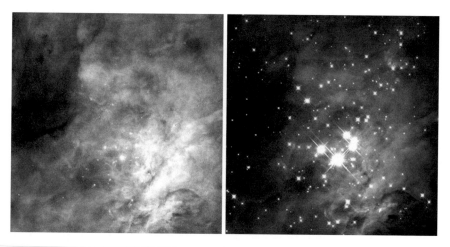

FIG. 1.6 The Trapesium in Orion (*Courtesy*: NASA)

a significant portion of the very stuff from which stellar systems are actually made is not something that one can easily understand. Where, it might be asked, did all these microbes come from in the first place?

Hoyle and Wickramasinghe have their answer to this question. They see life as arising in comets. Comets, as planetesimals, are the first bodies to form within collapsing clouds of interstellar material; material which (according to these researchers) largely consists of the desiccated corpses of earlier generations of microbes. But comets also possess another feature. They evaporate. At least, those that periodically come relatively close to their parent star evaporate and it is by virtue of this process of evaporation that they come to sport the fuzzy heads and tails that are characteristically associated with these bodies. The solid asteroid-like nucleus of the comet shrinks as its ices sublimate into surrounding space, in the process wafting away small solid particles which eventually leave the home stellar system and blend with the cosmic dust of the galaxy. If Hoyle and Wickramasinghe are right, many of these small solid particles are bacterial spore! Drifting through the void of interstellar space, these spores will eventually be swept into the galactic dust lanes and gathered together with their counterparts from other stellar systems. Some will eventually become incorporated into evolving pre-planetary nebula, and the whole process begins anew.

There is something aesthetically pleasing about this scenario, but that does not necessarily make it true and, indeed, very few astrophysicists have been enticed by its apparent beauty.

The thesis would gain a good deal more credibility if the structures found in certain meteorites truly are fossilized bacteria, as Hoyle, Wickramasinghe and their small but dedicated band of supporters claim. This is a radical claim and, as such, has drawn a great deal of criticism and alternate explanations. Of course, being radical does not make it incorrect, but it does mean that it will need to amass some very strong and water-tight evidence in its support if it is going to be accepted by the main stream of scientific thought. To date, the general consensus of opinion is that it has failed to accomplish this.

Moreover, even if these meteorite structures do turn out to be microfossils of bacteria or similar organisms, for the

Hoyle-Wickramasinghe scenario to work in its full, cosmic, setting it would be necessary to show that these microfossils are common to comets, not just in our own Solar System, but in essentially *all* Solar Systems.

The latter would probably be a fair assumption, at least for the Solar Systems of stars having broadly similar composition to that of the Sun. The composition of interstellar material seems broadly consistent across the galaxy, so it is a fair assumption that stars, planets and comets forming in different regions will have broadly similar compositions. The real discrepancy here would be between those systems forming now after generations of stars have enriched the interstellar medium with heavy elements and those early systems where such elements are depleted in comparison with solar values. This galactic evolutionary factor adds a complication which must be faced if this scenario is to be taken seriously.

The other immediate issue ultimately depends upon whether those meteorites in which microvesicles ("microfossils" as Hoyle-Wickramasinghe supporters would say!) are representative of the stuff from which comet nuclei are made. In short, are these meteorites from comets? In endeavoring to answer this, we will need to leave our main discussion for a while and take a look at the contentious issue of cometary meteorites.

The Case for Cometary Meteorites

There was a time, not so very long ago, when hardly anybody thought that any meteorite came from a comet. In fact, in 1975 a paper was published by meteorite expert E. Anders that appeared to rule out comets as a source for any meteorite examined thus far. In an ingenious piece of astronomical detective work, Anders used the ratio of solar-wind-implanted to cosmic-ray-produced noble gases in meteorites to infer the nature of the regolith of the parent body with respect to its distance from the Sun and the regolith-producing meteoroid impact rates that it endured. He concluded that both matched the distances at which asteroids are typically found. Ergo, all meteorites examined to that date apparently derived from asteroids.

The Anders' argument, though ingenious and superficially convincing, turns out to have an Achilles heel. A paper published by H. Campins and T. Swindle in 1988 pointed out that Anders' argument is convincing only if all comets are assumed to hale from the distant and sparsely-populated Oort Cloud, about which more will be said in the following chapter. However, the eccentric orbits of these objects means that with few exceptions, any potential meteorite that might come from one of them would meet the Earth's atmosphere at such high velocities as to be reduced to dust before reaching the ground. If any cometary meteorites do lurk in our collections, the chances are that they were produced by short-period comets of the Jupiter family. A few of these are known to shed meteoroids into our atmosphere at sufficiently low velocities to fall as meteorites if any are large enough and strong enough to make it through the planet's ocean of air. Comets of this class are now believed to have ventured inward, not from the Oort Cloud, but from the far closer Kuiper Belt where the chance of impacts by meteoroids and dust particles is thought to be about as high as it is within the asteroid belt. Also, Campins and Swindle argue, the uncertainties in the solar-wind-implanted gas concentrations and surface residence times used by Anders were wide enough to cover both the Kuiper and asteroid belts, though not the Oort Cloud. The upshot of this is that, in the opinion of Campins and Swindle, Anders has not ruled out a cometary origin for some meteorites. He has only excluded Oort Cloud comets as the origin for known meteorites. Even here though, we cannot say that comets coming in from the Oort Cloud are not capable of producing meteorites per se or that no such meteorite has ever fallen, only that they must be rare and that none has been collected and examined. As already remarked however, the high velocity of most meteoroids from this class of comets probably accounts quite well for this rarity.

In the opinion of most experts who have studied this question, if cometary meteorites do exist, they are likely to be found amongst the classes of Type 1 and Type II carbonaceous chondrites. These most nearly correspond to the chemical composition of comets and both classes of object are thought to come nearest to representing the primitive composition of the pre-solar nebula itself. Fortunately for Hoyle and Wickramasinghe, these are the two classes where most of the microfossil-like structures have been found. Recent discoveries concerning the nature of comets and asteroids suggest that these two classes of object merge into one another rather than (as previously thought) exist as two distinct populations where never the twain shall meet. In view of this, it may be that some meteorites of these two classes come from objects that we would categorize as asteroids while others hale from indubitable comets. That, however, is another issue that will not be pursued here. If any of the meteorites residing in our laboratories and museums have come from comets, then we have already sampled cometary material and discovered that it contains not simply interesting organic substances but equally intriguing organic structures as well. Considering the probable ubiquity of cometary bodies in the wider universe, both the organic material and organic structures found in meteorites such as Murchison might then represent something very widespread indeed throughout the universe.

If Hoyle, Wickramasinghe and their colleagues are even partially correct in their seemingly weird and wild speculations, that something might just be definable as life. But, though there are arguments that some carbonaceous meteorites might be from comets, is there any real evidence to support this? Actually, there is. The Taurid meteor complex, associated with Comet Encke, occasionally drops a fireball with the characteristics of a meteorite, though none has been uncontroversially recovered.

The meteorite most often cited as a possible Taurid is the Maribo, a CM2 (Type II) carbonaceous chondrite that fell in Denmark on 2009 January 17. A study by H. Haack et al. and published in 2011 revealed the orbit of this body to be consistent with it having been an "outlier" of the Taurid meteor complex. In fact, its orbit was of a remarkably similar *type* to that of Encke itself, although the two orbits were not all that alike in respect to their specific parameters. Asteroidal fragments can, however, find their way into Taurid-like orbits, possibly raising some question marks over the object's parentage, although its carbonaceous nature might be seen as supporting an Encke connection.

The cosmic ray exposure age of this meteorite is about 1.1 million years, plus or minus around three hundred thousand years. It is unlikely that Encke has been in an orbit similar to its contemporary one for so long a time and the age of the Taurid complex is generally supposed to be only between 20 and 30 thousand years.

Nevertheless, the lump of matter that became the Maribo meteorite may have been exposed on the surface of Comet Encke, possibly following a collision between the comet and a meteorite, at a time when the comet was a dormant object, maybe following a Centaur-like orbit in the outer Solar System.

Haack and colleagues concluded from this work that all CM2 meteorites probably have their origin in Encke and the Taurid stream. This is, I think, going too far. If this meteorite truly is a fragment of Encke, all we can really say is that some meteorites of this type originate in *some* comets. If Encke is the parent of at least one meteorite, other comets may be the parents of similar bodies and, indeed, it will be argued below that one famous meteorite may have Comet Finlay as its parent. It might be suggested that the lion's share of comets of very short period have a lineage that can be traced back to a common ancestor; a very large comet that became trapped within the inner planetary system many thousands of years ago and subsequently broke up progressively into a myriad of smaller objects, some of which remain visible as the short-period objects of today. If several of these later-generation comets are parents of carbonaceous meteorites, these latter can in that respect be said to, ultimately, have a common origin. This speculation will be raised again in Chap. 4 of this book.

Before leaving the question of Maribo's origins, it is also worthwhile noting that its orbit bears a striking resemblance to that of the Apollo asteroid 1991 AQ (=1994 RD = 85182). According to J. Drummond's D' criterion for distinguishing associated orbits, a D' value of 0.105 or less implies association. The D' criterion for the Maribo and asteroid 85182 orbits is just 0.04. It is interesting to note that this asteroid, together with four other similar bodies including Hephaistos is suspected by several astronomers of association with the single-apparition periodic comet Helfenzrieder, observed as a rather bright naked-eye object back in 1766, clearly during an anomalous surge in brightness and activity. In studies published in *The Observatory* in 1994 and in *Earth, Moon and Planets* the following year, D. Asher and D.I. Steel argued that this association, like the Taurid complex itself, most likely arose from the disruption of a large periodic comet several thousands of years ago and that this hypothetical comet may itself have broken away from the Taurid complex progenitor (a considerably enlarged Encke) at a still earlier date. In a sense, the entire Hephaistos family is, according to these researchers, an outlier of the Taurid complex, mainly distinguishable from the former by the difference in the longitude of perihelion (orbital orientation) between the respective members of each group. If that is correct, the Maribo meteorite may be related to two comets and several Apollo asteroids, the latter most likely also being defunct comets. (See Appendix A for further details.)

For meteorites to fall from cometary meteor showers, the velocity at which shower members enter Earth's atmosphere must be low. Until recently, it was frequently stated that speeds of under 20 km (about 12.4 miles) per second (slow by meteor standards) was required, yet the Maribo meteorite entered Earth's atmosphere having a velocity of 28.5 km (just under 17.7 miles) per second and on 2012 April 22 another meteorite fell at Sutter's Mill in California after having entered our atmosphere at 28.6 km/s. The latter was also a CM2. It is interesting that both Sutter's Mill and Maribo, the meteorites having the highest atmospheric velocities recorded to date, belonged to this second most fragile meteorite class. If these meteorites could survive such velocities and arrive at the planet's surface as something more than wafts of fine dust, then surely other more robust objects could also survive.

Having raised the subject of the Sutter's Mill meteorite, it is worthwhile to note that the orbit of this meteorite has the right perihelion distance, eccentricity and orientation for membership of the broader Taurid complex, although there is no close association with specific Taurid asteroid orbits. The orbit does, however, reveal some degree of match to the May Arietid meteors. Moreover, this meteor stream also displays a certain orbital match with the southern branch of the Taurid stream. This may be something worthy of further investigation (see Appendix B).

Several other meteor showers of cometary parentage should also be considered as serious contenders for possible meteorites. Two such meteor streams from which meteorites are suspected to have fallen are the Pons-Winneckids and the Omicron Draconids. The first of these consists of a cluster of related radiants straddling the constellations of Draco and Bootes and active from the latter days of June until well into the following month, but most apparent around the very end of June and the first few days of July.

As their name implies, they arise from debris spread abroad by the short-period comet Pons-Winnecke. This shower is also known as the Bootids and some old meteor catalogues list it as the Iota Draconids. The second shower mentioned above is active during the latter half of July and is widely believed to have been spawned by the comet 1919 Q2 (Metcalf). This relationship is interesting, as the orbit of this comet implies that it was from the Oort Cloud and probably making its first trip to the inner Solar System in 1919. Indeed, its maiden trip might also have been its last as planetary perturbations caused its orbit to become slightly hyperbolic as it retreated back into deep space. A small mystery remains however, in so far as it is possible that meteors from this stream were observed prior to 1919; at least, meteors were observed in earlier years which appeared to radiate from the same general region of sky. Maybe the comet was not making its very first trip to the Sun (though its inward orbit was so elongated that any prior passage must have been millions of years earlier) or maybe the comet split whilst still far out in space and the other fragment reached perihelion, unobserved, decades before 1919. Or, maybe, the pre-1919 meteors were not Omicron Draconids after all.

In any case, in the year 2008, on the night of July11, a brilliant member of the Omicron Draconid stream lit up the heavens over Spain and was well recorded on cameras of the Spanish Meteor Network. The meteoroid is calculated to have been about 1 m in diameter and to have possessed a mass of approximately 2 tons. This fireball possessed the characteristics of a meteorite-dropper but unfortunately no fragments were recovered. This is doubly regrettable, as this seems to have been a meteorite derived, not just from a comet, but from an Oort Cloud comet of very long period. It would surely have been a prized possession if found.

Like the Omicron Draconids, the Pons-Winnecke stream is also a very weak one, albeit one that can (and occasionally does) spring a surprise. This happened in 1916 when the shower was unusually active and it has happened several times since then. On other occasions, the stream has surprised observers not by producing larger-than-normal numbers of meteors but by giving rise to one of unusual brilliance. Sometimes, a very brilliant fireball issues forth from the Pons-Winnecke radiant. The prime example was the one seen from Tajikistan on the night of 2008 July 23 which, like the Spanish Omicron Draconid earlier that same month, possessed all the characteristics of a meteorite. Assuming that these fireballs were bona fide members of the respective meteor streams, it seems that fragments of comets Metcalf and Pons-Winnecke ended up on Earth that year, but, alas, neither was recovered.

From the behavior of these fireballs, N.A. Kanovalova and colleagues deduced a strength comparable to that of typical Type I or II carbonaceous chondrites. In a paper presented at the 43rd. Lunar and Planetary Science Conference (2012), these authors also gave details of some earlier Omicron Draconid and Pons-Winneckid meteors, revealing quite a large range of tensile strengths displayed by these bodies. A member of the first stream observed on July 12, 1959 and one of the second shower observed on June 23, 2004 were weaker than the most fragile meteorites and, even had they been sufficiently large, would not have survived their passage through our planet's atmosphere. On the other hand, an Omicron Draconid fireball recorded on July 14, 2007 appeared strong enough, but was probably too small to have fallen as a meteorite. Clearly, meteoroid strength differs even between members of the same stream; something observed too frequently to be seriously doubted.

Some 29 years before the possible cometary meteorites of 2008, a very interesting event occurred near the town of Allan in Saskatchewan, Canada. On October 19, 1979, a slow fireball on a low trajectory glided across the skies and was well observed photographically be a fireball-monitoring station. The radiant of this meteor was found to lie close to the star Eta Ophiuchi. The fireball is listed as Number 498 in the April 1989 paper by Ian Halliday et al. on unrecovered Canadian meteorites photographed by the Canadian fireball network between the years 1971 and 1985. This fireball had the characteristics of an event that probably produced a small meteorite. Moreover, the fragmentary tendency of the meteoroid was interpreted by Halliday as indicative of a carbonaceous chondrite; the only one in his unrecovered list. From the trajectory of the fireball, plus its fragmentary nature and the low elevation of its radiant as seen from that part of the world, it is likely that any mass reaching the ground arrived in the form of many small stones distributed along a long and narrow strewnfield crossing agricultural country on the edge of hilly terrain. It is a pity that the

presumed meteorite was not found, but in view of the circumstances of the fall and the nature of the object itself (a fragile carbonaceous chondrite), this is not too surprising. Fortunately though,
the good photographic record of its atmospheric flight enabled
both its radiant and, indeed, the orbit itself to be calculated.

In their 1988 paper, Campins and Swindle concluded that
this object was most likely a fragment of the dormant comet
Wilson-Harrington. The story of this object is an interesting one.
On 1979 November 15, E. Helin and S.J. Bus discovered a fast-
moving "asteroid" on a photograph taken with the 46-cm Schmidt
telescope at Palomar Observatory. The object was clearly a near-
Earth interloper and was given the provisional designation of 1979
VA. After the orbit was computed, two interesting facts emerged.
First, the orbit was remarkably like that of a short-period comet.
Secondly, the object had made an Earth-grazing encounter prior to
its discovery, having passed our planet at a distance of just 0.091
Astronomical Units (AU; a distance equivalent to the average
radius of Earth's orbit around the Sun) on October 29. Interestingly,
reflectance spectra of the apparent asteroid suggested a color rather
similar to that of carbonaceous meteorites.

After a sufficient number of observations had been amassed
to allow an accurate orbit to be calculated, searches were made
of older survey plates in the hope of finding images from former
returns. This is where things became really interesting. In 1992,
E. Bowell of Lowell Observatory succeeded in finding earlier
images of 1979 VA; on photographs taken as part of the National
Geographic Society—Palomar Sky Survey back in 1949. These
images had already been recognized, but as those of a comet not an
asteroid! The object was listed as Comet Wilson-Harrington and
was suspected of being periodic, although the observational arc
was too short for a reliable elliptical orbit to be computed (Fig. 1.7).

Back in 1949, images were obtained on only five nights and
on all but two of these, the comet appeared asteroidal. On the
nights of November 19 and 22 however, a small tail was noted
that was more clearly visible on the blue plates. Following the
identification of Wilson-Harrington with 1979 VA, these images
were enhanced to make sure that the "tail" was not simply a photographic defect or something equally as uninteresting. But the
tail was real! Moreover, its blue color and short duration indicated
that it was gaseous and not simply a plume of dust kicked up

FIG. 1.7 Comet Wilson-Harrington in 1949 (*Courtesy*: ESO & Palomar Observatory)

by the impact of a meteorite on the surface of an inert asteroid. According to a 1997 study by Y. Fernandez and colleagues, the tail most probably consisted of water and carbon monoxide ions; just the species that we should expect to find in the ion tail of a comet. It is not at all clear why it was not visible on the other nights that the comet was photographed, however it is worth noting that the gas tails of other comets have been seen to experience periods of brief intensification at times. These events appear to owe more to solar influences than to anything intrinsic in the comets themselves and it is possible that Wilson-Harrington possessed a tail too weak to be observed for most of the 1949 apparition, but was triggered by solar activity to flare for a brief time and, in so doing, betrayed the true nature of its host.

It is not known if the comet continued to display regular, albeit very weak, activity between 1949 and 1979 or whether the burst of activity in 1949 was an anomalous event in a comet that had already become dormant at most returns. In either case, the comet's dormancy is probably due to an accumulation of non-volatile material on the surface shielding the deeper ices from the Sun's warmth rather than an indication that its store of ices has become completely exhausted. It is interesting to note that the comet came even closer to Earth on 1919 September 30 than it did in 1979 but the 1979 approach will stand until 2198 September 11, although it will pass only marginally outside the 1979 distance on 2155 October 30.

In all this talk of close approach however, the reader will undoubtedly have noticed that the 1979 minimum distance occurred just 10 days after Halliday's fireball Number 498.

Mention should be made at this juncture of an apparently non-cometary meteorite that nevertheless happened to fall during one of the most spectacular meteor storms of the past two centuries. This was the Mazapil meteorite of November 27, 1885, the fall of which coincided with the peak of the great Andromedid storm spawned by debris from the now-defunct comet Biela. The Andromedid meteors are slow, implying that this stream possibly could produce a meteorite. The problem is, Mazapil is an iron meteorite; hardly the sort of object that one might expect to have formed within the nucleus of a small comet! Assuming that Biela had not once been part of a very much larger object within which conditions more conducive to the formation of iron meteorites were possible, the most likely explanation for this coincidence of meteor storm and meteorite is exactly that—coincidence. On the other hand, there remains a very small probability that the meteorite was an asteroidal fragment that the comet had picked up at some time in the past and subsequently released into the developing meteor stream. This process of asteroidal meteorites hitchhiking on comets was initially proposed by E. Opik, not specifically for this object, but as a general mechanism for delivering meteorites to Earth's vicinity. Opik's proposal, at that time, seemed to be the most efficient way of bringing asteroidal fragments into Earth's region of space, however subsequent research has found that such a mechanism is not necessary and the idea has now been abandoned as the principal means of delivering meteorites to our planet. That said however, there is no reason why it should not happen on rare occasions and it is possible (though, we admit, very unlikely) that the Mazapil meteorite was one such occasion. Even if that is true though, this meteorite could still not be claimed as a truly cometary one, as its

Project 1.1: Meteors from a Sleeping Comet?

Even though there will be no close approaches by Wilson-Harrington itself for over a century, Earth continues to pass within 0.05 AU of its orbit on two occasions every year. The first approach occurs on 8 September and the second on 2 October, according to computations by J. Drummond and published in the year 2000 (Icarus 146, pp. 453–475). It was from this second radiant that the unrecovered meteorite, Halliday's fireball Number 498, apparently issued.

The two radiants given by Drummond are;

Date	RA (2000)	Dec. (2000)
September 8	18 h 41.22 min	−24.8°
October 2	17 h 23.58 min	−21.7°

Is there an annual meteor shower from either or both of these radiants or has Wilson-Harrington been inactive for so long that any meteors shed by this object during its presumably more active youth have long since drifted into the sporadic background?

The most spectacular evidence that any particles still follow the comet appears to be fireball 498 although this object has been suspected of giving rise to some meteor activity radiating from Sagittarius during September.

Meteors can arrive from Wilson-Harrington throughout September and most of October, but any activity should peak around the dates mentioned. Remember that the peak (if one occurs at all) will be a very low one.

Experienced meteor watchers might like to monitor the radiant positions on and around the given dates. Due to the likely faintness of any members of this system, a watch by several meteor observers located under a truly dark sky is the best option. While positive results would be exciting, high quality negative results would also be interesting as these can set a maximum on the number of meteors that may still be associated with this dormant object and this in turn should help to estimate the length of time that has elapsed since Wilson-Harrington was a fully active comet.

properties surely place its origin within a differentiated asteroidal body.

The Murchison Meteorite—Out of the Silent Comet?

Back in the late 1980s, the author became interested in the circumstances surrounding the fall of Australia's most famous meteorite and actively gathered quite a number of accounts of its appearance and trajectory. To make a long story short, this information enabled an approximate but relatively satisfactory radiant position to be determined and this in turn led to a reasonable calculation of its orbit by I. Halliday and B. Mackintosh.

Several objects were proposed as its possible parent. One of these was none other than Wilson-Harrington, which is very interesting in view of Halliday's Canadian fireball. Another proposed possibility was the near-Earth asteroid 1989 VB. There was a report that a 1990 observation of this object showed it to be C-Type and therefore a good prospect as the parent of a carbonaceous chondrite, but it seems that this report was not correct. According to J. Drunmmond, there appears to be a dynamical association between 1989 VB and the dark asteroid (2061) Anza, but whether this implies that the former is a fragment of the latter (and therefore of the same type) or whether it simply means that each experienced similar dynamical influences is not known. However, even if 1989 VB is a fragment of Anza, the reflectance spectra of the latter does not closely match that of meteorites of the Murchison variety. For one thing, bodies with spectra of this type are thought to be anhydrous, in contrast to Murchison which has quite a high water content of 12 %. Also, the radiant of the Murchison meteorite, as determined by Halliday and Mackintosh, lies somewhat further south than those of 1989 VB, Anza and Wilson-Harrington. It does, on the other hand, lie very close to a series of radiants calculated for theoretical meteors from the periodic comet Finlay (Fig. 1.8).

Finlay is a strange comet in several ways. In 1981, Drummond noted that all but four of the comets of short period, known at that time, that passed within 0.08 AU of Earth's orbit (a total of 28) had associated meteor showers. Since then, meteor displays have either been confirmed or strongly suspected for three of Drummond's silent comets. The odd one out is Finlay. An examination as to why Finlay should be the remaining silent comet

FIG. 1.8 Comet Finlay remotely images from New Mexico January 17, 2015 (Copyright Michael Mattiazo, used with permission)

was later conducted by P. Beech, S. Nikalova and J. Jones with some surprising results. These researchers found that most of the meteoroids shed by this comet should be quickly perturbed by the gravitational influence of Jupiter until they either stayed outside Earth's orbit altogether or else had their intersection of our planet's path delayed until around 3 months after the Earth passed through the place of crossing. Either way, very few Finlay meteors should appear in our skies. Moreover, these authors also draw attention to the fact that Finlay's activity seems somewhat erratic. Maybe, they suggest, this comet spends most of its existence in a dormant state like Wilson-Harrington and is only awakened from time to time by meteoroid impacts breaking away some of its insulating crust as it passes through the asteroid belt on its very low-inclination orbit. This would imply that it is, at best, a feeble producer of meteoroids. The authors suggest that the absence of observations of this comet during its pre-discovery close encounters with Earth in 1714 and 1827 indicate that it was probably dormant or at least semi-dormant then. The brightest return was that of 1906 when it was very well placed and became brighter than magnitude seven. It appears to have faded steadily throughout the rest of the twentieth century, until its return of 2008 when it surprisingly became bright enough for observing with small telescopes. Due again at the end of 2014, it remained very faint until several weeks prior to perihelion, when it rapidly surged in brightness and attained an even greater luster than it had achieved in 2008. But the real surprise occurred in January 2015 when it experienced a strong outburst and became intrinsically brighter than at any previously observed return. Its apparent brightness almost equaled that of 1906, even though it remained further from Earth than at the earlier apparition. It certainly shed dust—and presumably meteoroids—in 2015. This comet will come to within 0.05 AU of Earth on 27 October 2060 and in the unlikely event that minimum approach coincides with a flare to the January 2015 levels, the comet would then brighten to around zero magnitude.

It would be nice if there was a recognized Finlay meteor shower sharing the radiant of Murchison, but even if most of the comet's meteoroids are perturbed out of Earth's way, that does not mean that an occasional one might still reach us. Moreover, the Beech et al analysis was based on very small particles ejected at relatively high velocities from the comet whilst near perihelion. Larger pieces that may break away anywhere along its orbit might be a different matter.

A problem that has been raised concerning a Finlay origin of Murchison is the determined cosmic-ray exposure age of the meteorite. This reveals that the parcel of matter that became the Murchison has been exposed to a space environment for around two million years and it has correctly been argued that Finlay could not possibly have been in its present orbit for such a long interval. My suggestion, however, is that the Murchison meteorite may have been tossed to the surface of the comet by an impact while Finlay was still a member of the Kuiper Belt. It would then have been exposed to the space environment, albeit not as a free-floating object, until the comet dynamically evolved into its present orbit and the shrinking of its nucleus through the erosive effects of cometary activity caused the meteorite to break loose. This final breaking away from the parent might have been very recent indeed.

Another difficulty raised against a cometary origin of Murchison (or any meteorite for that matter) is the supposed inability of most comets to provide the conditions of heat and pressure that even the least processed of meteorites have evidently encountered whilst inside their parent bodies. Although Murchison has not experienced great heating and pressure, it has obviously experienced a mild degree of both and questions have been raised as to whether something as small and icy as the nucleus of an average comet can provide such conditions. For many comets, the answer is probably in the negative, however it is now generally agreed that most short-period objects such as Finlay were once denizens of the Kuiper Belt and that collisions between Kuiper Belt members occur from time to time. We also know that some Kuiper Belt inhabitants are rather large. The dwarf planet Pluto is just one example and there must also be great numbers of objects of similar size to the larger asteroids that are found within the main asteroid belt between Mars and Jupiter. Like the denizens of this nearer belt, many members of the Kuiper Belt are presumably fragments of far larger objects that have been broken apart during collisions. It may be that the majority of comets of short period are actually fragments of these larger bodies. Consequently, although it might be true that an object the size of Finlay (probably about 1 or 2 km in diameter) cannot maintain conditions under which meteorites like Murchison can form, the same probably cannot be said for the larger bodies of which at least some Finlay-sized comets are but the fragments. Add the presence of short-lived isotopes (for example, AL 26) that are known to have been present during the Solar System's youth, and we cannot rule out the presence of relatively highly metamorphosed meteorites from the Kuiper Belt. Further consideration of this possibility would, however, take us too far from our topic.

Table 1.1A lists several theoretical radiants of meteors released from Comet Finlay during a number of its perihelion passages. Most of the radiants are from calculations by P. Beech et al. although some computations by A. Verveer have also been included as well. Table 1.1B gives two possible radiant positions for the Murchison meteorite as computed by Halliday and McIntosh. The first radiant is for their preferred orbital solution, assuming an initial velocity of 13 km/s (8.06 miles/s) and giving an eccen-

Table 1.1

(A) Positions of Theoretical Radiants of Finlay Meteors			
Finlay perihelion year	RA	Dec.	Date
1585	19 h 43 min	−50° 23′	Sep. 28
1721	20 h 21 min	−61° 33′	Sep. 17
1886	17 h 50 min	−35° 22′	Oct. 16
	17 h 39 min	−35°	Oct. 15
1893	17 h 43 min	−34° 57′	Oct. 19
1906	17 h 31 min	−33° 47′	Oct. 27
1919	17 h 47 min	−37° 27′	Oct. 8
	17 h 39 min	−37°	Oct. 8
1926	18 h 16 min	−39° 16′	Oct. 3
	18 h 44 min	−39°	Oct. 1
1953	18 h 11 min	−38° 56′	Oct. 4
1960	18 h 02 min	−40° 47′	Sep. 29
	17 h 55 min	−41°	Sep. 28
(B) Murchison Meteorite Possible Radiant (1969 Sep. 28)			
Assumed initial velocity	RA	Dec.	
13 km/s	19 h 7 min	−37°	
14 km/s	18 h 47 min	−41°	

tricity of 0.487, an inclination of 2.5° and a perihelion distance of 1.0008 AU. The second assumes an initial velocity of 14 km/s (8.7 miles/s) which yields an eccentricity of 0.628, an orbital inclination of 3.7° and a perihelion distance of 1.0016 AU. Both solutions assume that the meteorite came in at an elevation of close to 25°, as derived from information collected by the present writer. The first solution was preferred mainly because the resulting aphelion distance (2.9 AU) is near the average of meteorite orbits available at the time. However, more recent orbital derivations of carbonaceous meteorites have yielded aphelia from around 3.2 AU to greater than 5 AU, suggesting that the second and more eccentric orbit (with an aphelion distance of 4.38 AU) might in truth be the preferred one for a carbonaceous chondrite. In any case, the range of possible entry velocities nicely matches those computed for Finlay meteors. If there are such entities as Finlay meteors, they would be amongst the slowest of all cometary meteors, which would be good news for the survival—as meteorites—of any especially large ones.

Considering the inevitable uncertainties involved in the determination of the Murchison orbit and radiant, the comparison with the theoretical Finlay results is very interesting indeed.

It is also interesting to note that some of the possible Murchison orbits, that is to say, those within the range of likely orbits given by Halliday and Mackintosh, indicate that the meteorite and comet may have been pretty close together around the time of the unobserved Finlay perihelion in late April 1940. It is tempting to suggest that Murchison may have split away from the comet's nucleus around that time, although given the uncertainties in the meteorite's orbit and the consequent uncertainty of its real position at that time, this is probably just a piece of useless speculation.

Project 1.2: Is Comet Finlay Completely Silent?

Experienced meteor watchers might like to look closely at the region of sky around the radiants in Table 1.1 just in case a few Finlay meteors do manage to encounter our planet. Any meteors from this system will be very slow and therefore fainter than those caused by particles of equivalent size belonging to most other cometary meteor streams.

Although no Finlay meteors have been proven to exist, there have been some possible sightings and confirmation of even slight activity (as well as good quality negative results) would be beneficial and should be of interest to meteor watchers.

Amongst possible sightings, the Western Australian Meteor Section observed a few very slow meteors during the 1980s and derived possible radiants near 18 h 8 min and −40° from September 25–28 and around 18 h −40° from September 29–October 2. Activity was so low however, that the observers suspected that these radiants might be nothing more significant than random alignments of unrelated sporadic meteors. Nevertheless, the objects concerned appeared to share the same characteristics of being very slow, reddish in color and readily fragmenting which might imply a genuine association.

A watch in the late 1980s by the author produced a handful of meteors that, with one exception, were either travelling in the wrong direction, at the wrong velocity, or both. The exception was a nice orange-colored slow meteor of about magnitude zero which appeared consistent with something issuing from one of the Finlay radiants. A lone meteor seen by one person who has never been a serious observer of meteors is not, however, good evidence for the existence of a meteor stream. Quite possibly, this was nothing more than a fortuitously placed sporadic.

More significant is the result of search for possible meteor/parent body associations conducted by J. Greaves and published in *Radiant*, 2000. Using the data-base of the Dutch Meteor Society and employing Drummond's D' criterion, Greaves found one meteor apparently related to Finlay and three more that may have been related either to Finlay or to the periodic comet Denning-Fujikawa. One member of this trio was slightly closer to Finlay whereas the other two were a little closer to Denning-Fujikawa.

Before leaving the subject of a possible Finlay/Murchison connection, mention should be made of a paper by A. Terentjeva and S. Barabanov published in the International Meteor Organization journal *WGN* for August 2011. These authors argue for an association of objects related to Comet Finlay. This association appears to have both a northern and a southern branch and appears to be a real physical system rather than a mere random grouping of unrelated bodies. The northern branch consists of the comet Haneda-Campos, together with asteroids 2061 Anza and 2001 PE1, while the southern branch consists of Comets Finlay and Wilson-Harrington along with the September Gamma Sagittarid fireball stream, asteroids 1997 YM3 and 2000 PF5 together with the Murchison meteorite. To the northern branch we might also add the October Capricornid meteors (maximum on October 4) which may be debris from Haneda-Campos, and to the southern branch, the asteroid 1989 VB and Halliday's unrecovered Canadian meteorite No. 498. There is, indeed, a strong association between the September Gamma Sagittarid orbit and that of 1989 VB, so it is possible that these meteors are debris from this asteroid. (It should be mentioned that Wilson-Harrington has also been suspected as this stream's parent body, although the association between the orbit of this body and that of the fireball stream is less than convincing.)

The October Capricornids apparently experienced a burst of activity in 1972 as noted on the night of October 2/3 of that year by Western Australian meteor observers, Derek Johns, Denis Rann and Michael Buhagiar who recorded a rather lively shower of bright meteors (from about magnitude 3 to –2, but mostly ranging between +1 and –2) diverging from that region of the sky. Haneda-Campos arrived at perihelion around October 20 in 1972 and approached the Earth to within 0.3 AU just 10 days later. However, as it still awaited discovery at the next return, nobody was aware of this fact at the time.

It is interesting to note that the comets within this group seem to be either dormant or heading in that direction and we are led to ponder whether some of the purported asteroids listed here are in reality comets which became dormant prior to their discoveries. As we have seen, Wilson-Harrington would not have been listed as a comet except for a brief and very weak bout of activity in 1949. Maybe one or two of those listed as asteroids had similar outbursts that we simply missed.

It is also interesting to note that Haneda-Campos has not been seen since its discovery apparition in 1978. The next predicted return late in 1984 was admittedly less favorable than those of 1972 and 1978, but had the comet been intrinsically as bright as it was at discovery, it should still have been an easy object for moderate-sized telescopes. In fact, together with the comets Encke and Crommelin, it was even proposed by the International Halley Watch as a practice comet for the 1986 return of Halley. Fortunately, this proposal was not acted upon and Comet Crommelin became the object of choice. Haneda-Campos failed to show up on schedule. It is possible that this object has either gone dormant or disintegrated since its discovery, but it is also possible that it had been dormant prior to 1978 and that year experienced an isolated burst of activity that helped bring about its discovery. Indeed, its behavior that year was a little odd and suggested that some type of outburst was taking place. It would not be at all surprising if someday one of the robotic search programs found a tiny "asteroid" that turned out to be none other than Haneda-Campos.

Comet Finlay, as already remarked, appears to have had an erratic history of activity and appeared to be on its way to a Wilson-Harrington-like dormant state before its phoenix-like behavior during the 2008 and 2014 returns. What it will do in the future remains to be seen.

Strangely, asteroid 1989 VB initially was thought to have displayed some cometary activity at the time of its discovery. The discovery image of this object showed what appeared to be a diffuse glow just touching the asteroid, superficially appearing as a weak cometary coma. This is now thought to have been nothing more than a photographic defect, as a second, albeit weaker, image was also detected on another part of the photograph.

The meaning of the Finlay association, assuming that it is real, is unclear but it may be the result of a large cometary object disrupting in the relatively recent past, perhaps as a consequence of a close approach to Jupiter, leaving fragments large and small in its wake, some of which have been captured by this planet into the orbits of short-period comets. Such considerations however take us too far beyond the subject of this book.

What, therefore, can we say concerning the proposed Murchison/Finlay association? In short, while the evidence favoring it is intriguing, it falls short of actual proof. If this were a legal matter, one could safely say that there would be sufficient evidence to bring the case to trial, though probably not enough to secure a conviction.

So where does all of this leave the hypothesis that primitive organisms originate in comets, are returned again to the interstellar medium through cometary activity, desiccate into cosmic dust and in that form play a vital role in the birth of stars and planetary systems?

Sadly for this hypothesis, it must be said that even if the microstructures found in Murchison and similar meteorites are fossils of once-living organisms and even if these meteorites are from comets and do indeed represent typical samples of cometary matter, the hypothesis still lacks the proof required for such a radical conjecture. Summing up the situation, the most that can be said without further evidence, is that organic material is abundant in the universe and that the stuff of living organisms is also common to the nurseries of stars. There is, in the opinion of the present writer, also sufficient evidence that something like proto-cells have been found in carbonaceous meteorites and that some of these meteorites originate in comets. From this it follows that these proto-cell-like microvesicles most likely exist in at least some comets. Almost certainly, comets exist in stupendous numbers throughout the universe and a small percentage of these (which still amounts to an enormous number) approach their parent stars closely enough to partially evaporate and, presumably, expel quantities of their microvesicles in the process. If this line of reasoning is correct, there should be proto-cell-like microvesicles floating around within the interstellar medium. Even going this far will no doubt be seen as too radical for many people, but it is something that is at least worthy of further consideration. To that degree, Hoyle and Wickramasinghe might be correct.

Nevertheless, the aspect of the Hoyle-Wickramasinghe position that grabbed public attention more than any other is a good deal weirder and makes the above suggestion seem very cautious and conservative by comparison. Let us now take a closer look at this very controversial hypothesis.

Diseases from Outer Space

This sounds like the title of some third-rate science fiction horror movie, but it is actually a serious proposal made by one of the previous century's most eminent scientists. The scientist in question was, not surprisingly, Sir Fred Hoyle about whom we have already been speaking. Together with his colleague Chandra Wickramasinghe, he took the idea of panspermia one step further than even its most ardent supporters are normally willing to go. What these scientists proposed was the idea that comets are literally the breeding grounds of many of the diseases that infect life on Earth. Not that this is wholly a bad thing, as they also took the view that disease plays a positive role in biological evolution (an idea not unique to them actually), so that the occurrence of outbreaks of disease is really a double-edged sword.

The idea that extraterrestrial microorganisms may be harmful to life on Earth is not, in itself, to be dismissed entirely. Although the possibility is remote, it is not zero. Older readers will remember how the first Apollo astronauts to return from the Moon had to remain in air-locked isolation for a time after returning to Earth, just in case the Moon harbored some unfriendly alien bugs. While it is doubtful that anybody seriously believed in lunar pathogens, as a small possibility it could not be ruled out, so the planners of these missions wisely decided to play it safe. Bacteria, or something like bacteria, may inhabit the interior of comet nuclei and *may* be carried into Earth's atmosphere by interplanetary dust or even spread from the interiors of exploding meteoroids. Some may even arrive within carbonaceous meteorites. The possibility of any of these statements being true is tiny, but a slight theoretical possibility remains. Maybe the chances were greater in the early epochs of the Solar System, but even today it cannot be totally ruled out.

So certain bacteria may at least in theory arrive from an alien source, and a percentage of these may be detrimental to the health of terrestrial organisms, just as a small percentage of home-grown terrestrial bacteria are harmful, though most are not, despite what the manufacturers of household disinfectants would like us to believe. But then Hoyle and Wickramasinghe went a step further. They put forward the hypothesis that some of the supposed extraterrestrial pathogens were not bacteria but viruses.

It began to look as if Earth was under alien attack; not by bug-eyed monsters but simply by bugs. Indeed, Hoyle and Wickramasinghe even opined that the reason why most animals (humans included) have noses with nostrils pointing downward toward the ground rather than upward to the sky is because upward-pointing noses would make their owners more prone to infection from the constant rain of pathogens floating downward through the atmosphere from outer space.

One of the diseases singled out as a likely extraterrestrial contaminant was influenza; especially the virulent pandemic varieties that from time to time sweep our planet. Traditionally, pandemics begin in one region where a less virulent strain of influenza virus mutates into a more malignant form. Because human beings have not encountered this variety previously, we have little natural resistance and quickly succumb. The virus multiplies and is carried both on the wind and from person to person until the outbreak becomes widespread.

The most deadly flu outbreak in relatively recent history was that of 1918–1919. During those years, more people died of influenza than were killed in the four preceding years of what had been, until that time, the most deadly war in history. Although traditionally called Spanish Flu, the epidemic did not actually start in Spain. Indeed, it did not appear to start in any one spot in particular, and it is this very fact that Hoyle and Wickramasinghe pounce upon as evidence that something more than the traditional picture of epidemics was involved here. They draw attention to the fact that the first instances of the disease occurred on the same day in Bombay, India, and far away in Boston, USA. Yet, confining our considerations to the USA for the moment, it required a further 6 weeks before the epidemic spread from Boston to New York. Rather than starting in just one place as a mutant strain and spreading to the rest of the world, the Spanish Flu seems to have started in several places more or less simultaneously and subsequently spread outward from these different centers at a relatively slow pace. But why should the influenza virus spontaneously mutate at the same time and in the same way in such widely dispersed regions? That is not an easy question to answer from within the traditional position; not, at least, according to the opinion of Hoyle and Wickramasinghe.

If this pandemic and others of a similar (fortunately mostly of a somewhat milder) nature were caused by new strains of the virus arriving from outer space, there would be no reason why cases should not appear simultaneously in widely diverse regions. If Earth sporadically becomes enshrouded in a cloud of flu germs, sudden world-wide pandemics would follow almost inevitably.

Notably, the authors found what appears to be a correlation between major flu outbreaks and times of sunspot maxima. At times of stronger solar wind, small (virus-sized) particles are more readily swept along through the Solar System, just as terrestrial dust is more easily raised and penetrates into our houses on a blustery day.

Another common virus for which these authors propose an extraterrestrial origin is whooping cough. Indeed, in this instance a specific astronomical object has been named as the source; the well-known comet of short period, 2P/Encke. The principal evidence for this association is an apparent correlation between outbreaks of this disease and perihelion passages of the comet. There does indeed seem to be an uncanny relationship between whooping cough and Encke's Comet. The author recalls an incident not long ago when, upon noting that Encke was soon predicted to pass perihelion, facetiously thought "I suppose there will be a whooping cough outbreak now" only to see on the news bulletins a day or so later that health warnings were being sent to parents of school-age children alerting them to a fresh outbreak of whooping cough sweeping through the schools.

Two comments need to be made however. First, correlation does not necessarily imply causation. We will return to this in a little while, but for now it will be sufficient to say that just because two or more series of events appear to happen in sync with one another, that does not necessarily mean that these events are causally related or even that they are associated in any way.

Second, although it might be possible for bacteria to thrive within a self-contained ecosystem inside the nucleus of a comet, viruses are a different matter. These entities must not be thought of as a more primitive form of life than bacteria. They are simpler, but not more primitive in so far as they depend for their existence upon the presence of forms of life of a more sophisticated nature than themselves. Influenza, for instance, could not exist without the presence of complex animal life. Indeed, some people debate whether viruses should be classified as forms of life at all as they lack the property, often numbered as one of the essential features of a living entity, to replicate themselves unless they are attached to another and far more complex organism. In one sense, the virus plus the organism being infected becomes the living organism. Or maybe "life" should be defined in a less rigid sense that does not necessitate the ability to replicate without attachment to another entity.

In any case, this philosophical question does not directly concern us. What it does imply is the inability for most viruses to replicate within the confines of a comet's nucleus. If bacteria exist there, it could be supposed that bacteriophages (viruses that attack bacteria) could survive in such situations, but what about viruses like influenza and whooping cough? Why indeed should they be there in the first place? How did they come to exist in a region of space lacking any suitable host? These questions are not easily answered by the theory's supporters.

Weird Correlations; Is the Encke/Whooping Cough Association Another One?

Strange correlations abound wherever we look. Pseudo-sciences such as astrology, pyramidology and the like thrive on them. At one time or another, not just pseudo-scientists but genuine scientists have come up with some pretty weird correlations indeed, such as the hypothesis that the economic trade cycle is influenced by the solar cycle. It has also been suggested (and alleged evidence supplied) that democratic societies are more likely to vote in less conservative governments during times of sunspot maximum.

Whilst the speculations of Hoyle and Wickramasinghe are not on a par with any form of pseudo-science, we must wonder if the return of Encke's Comet to perihelion has any more influence on a whooping cough outbreak in schools on Earth than a peak in sunspot activity has on the election to power of a democratic socialist government. Even if we were to grant that the comet is full of the viruses, its return to perihelion would not necessarily spread them across the Earth. Supposing that the solar wind wafts billions of virus spore from the nucleus into a virus tail (for want of a better term), the Earth would not be affected unless the Earth-Sun-Comet geometry was such that our planet actually passed through this tail. The rarity of known passages through regular comet tails shows just how unusual this suitable geometry must be. We passed through a comet tail in 1861, probably encountered the edge of the ion tail of Halley's Comet in 1910 and may have swept up a small number of ion-tail molecules during an encounter in 1975. Clearly, this is not a frequent happening!

Encke does not approach Earth very closely, even on the most favorable apparitions and even the meteor showers associated with this object (the Taurid complex) consist of particles expelled by the comet thousands of years ago. Particles leaving the comet today are not the ones that are currently reaching the Earth. This alone makes any genuine association with passages of the comet and simultaneously occurring outbreaks of whooping cough hard to fathom.

If a correlation between the comet's orbital period and whooping cough outbreaks is verified, the explanation is unlikely to involve any causal connection between the two. I am not a virologist, so anything that I say on this subject should not be taken too authoritatively, but I wonder if the periodicity of outbreaks is related to the time taken for the immune systems of people who have been exposed to the virus, but without actually contracting the disease, to once again become vulnerable. The period of the comet itself—just 3 years and 4 months—is more stable than that of most comets as Encke's orbit is confined to the inner Solar System and suffers little perturbation from Jupiter. If a cycle of immunity having a similar duration does exist and if by pure coincidence this managed to come into sync with the orbital period of Encke, a very good correlation would become established, without any hint of a causal association.

Sometimes what superficially looks like a correlation turns out to have a different explanation altogether. In the 1970s, the author was involved in the investigation of UFO reports, having always suspected that after all the known causes of sightings are eliminated (and these range from mistaken sightings of Venus, bright or unusual meteors, down to events as mundane as the flashing blue light on a passing police vehicle) a residual is left that probably includes one or more phenomena that have yet to be officially recognized. There is nothing strange in thinking that some things remain undiscovered. On the contrary, it would be strange if there were not undiscovered phenomena. After all, sprites, the high-altitude discharges from thunderstorms, were not discovered until 1989 and earlier reports of glows associated with thunderclouds were relegated to much the same no-man's-land as UFOs.

In any event, a discussion paper was presented in the mid-1970s which, on the face of it, looked very promising. The author of the paper plotted the times of reports against days of sunspot maxima and days before and after these times. That is to say, times of sightings were plotted according to whether they fell on day zero (a day of sunspot maximum), day minus one, minus two etc. (days prior to those of sunspot maxim) or day plus one, plus two and so forth (i.e. days following those of maximum). In the words of the paper's author, "the results speak for themselves". And so they did ... or appeared to do. The graph looked impressively like a child's drawing of a mountain peak with the summit directly over day zero. If the phenomenon that gave rise to UFO sightings was so strictly tied to days of sunspot maximum, maybe it was caused by the interaction of the solar wind and the atmosphere of Earth; a relative, perhaps, of the aurora?

However, doubts soon started to creep in. For one thing, the correlation existed between the number of sightings and days of highest sunspot number relative to the period at which the sightings were made. But data was taken from right across the solar cycle and there was no attempt to correlate the dates of sightings with absolute sunspot numbers or with the sunspot cycle per se. In other words, at the quiet phase of the solar cycle, a day of maximum sunspot number might be one where a single tiny spot was visible, while at the other end of the cycle, the Sun might be peppered with spots even on those days prior to or following the ones classified as having maximum sunspot number. The graph was clearly plotting the proverbial mixture of apples and oranges.

Secondly, the sunspot number goes up and down over a period of very few days. Whether the Sun is quiet or turbulent, days of maximum (which, remember, might by just one spot at very quiet times) occur so frequently that there are few days that are more than 48–72 h away from one of maximum. So defined, maxima occur so frequently that just about any event we might like to choose will have a high probability of falling on some day of sunspot maximum. Or, if not on that actual day, on the day or two immediately following or preceding it. There will be few events, say, 4 days either side of maximum for the simple reason that there are very few such days.

To demonstrate this, look at Fig. 1.7. This plots mentions of Kim Kardashian on the Channel 9 (New South Wales) Web page during December 2014 as plotted against that month's days of sunspot maxima. This example was deliberately frivolous, as I doubt that even the most ardent supporter of the idea that sunspots are behind just about everything that happens in the Solar System could believe that they somehow influence the publicity given to Ms Kardashian (Fig. 1.9).

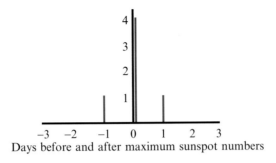

FIG. 1.9 An example of spurious correlation (*Courtesy*: The author)

Project 1.3: Spurious Correlations

Try this for yourself. Use a graph similar to that in Fig. 1.7 and plot just about anything you like against days of greatest sunspot number. No matter what you use, even random dates with no significance, it is almost certain that the outcome will look very similar to Fig. 1.7.

This little excursion is not intended to imply that any of the correlations claimed by Hoyle and Wickramasinghe are on the level of this "UFO" example. These scientists would have seen the flaw in that claimed "correlation" immediately. Yet, it is wise to remember how superficially good this example looked at first glance and to realize that not all spurious relationships are as obvious as this. Weird correlations have led to weird hypotheses in the past, and no doubt will do so in the future as well. To what extent that statement relates to the ideas of Hoyle and Wickramasinghe, the readers may decide for themselves.

2 The Birth of the Solar System: Some Unconventional Ideas

The Two Conventional Alternatives

The formation of the Solar System has seen many attempted explanations. Back in the 1700s, three gentlemen independently arrived at the nebular hypothesis of the Solar System's genesis. They were philosopher-astronomer Immanuel Kant, Pierre-Simon Laplace and Emmanuel Swedenborg. The scenario as proposed by these thinkers begins with a vast, hot, slowly-rotating nebula or cloud of gas and dust. As this cloud rotates, the outer regions start to cool and, in the process of time, the nebula slowly contracts and spins up through the principle of the conservation of angular momentum. As the velocity of rotation increased, rings of material were ejected from the main body in a process that has been likened, rather inelegantly, to mud being thrown off the rim of a rapidly spinning wheel. Eventually, the rings break up into discrete clouds of gas and dust which, with the process of more time, condense into solid orbs; the planets. Meanwhile, the bulk of the nebula contracts into the massive central body which becomes the Sun (Figs. 2.1, 2.2, and 2.3).

This hypothesis did a good job of accounting for two major features of the Solar System. First, according to this scenario, all of the planets orbit in the same direction and their orbits remain close to the plane of the Sun's equator. It also follows from this hypothesis that the direction in which the planets orbit is the same as that in which the Sun rotates about its axis. Nevertheless, as it stands, the nebula hypothesis fails to explain why the observed speed of the Sun's rotation is so slow. If it formed from a contracting cloud,

© Springer International Publishing Switzerland 2016
D. Seargent, *Weird Astronomical Theories of the Solar System and Beyond*, Astronomers' Universe, DOI 10.1007/978-3-319-25295-7_2

FIG. **2.1** Emmanuel Swedenborg (Portrait by Carl Frederik von Breda. *Courtesy*: Wikimedia)

it should have spun up dramatically (as the cloud was predicted to have done) from the conserving of angular momentum. As a ballerina spins faster as she folds her arms, contraction of any rotating mass must inevitably increase the velocity of rotation of that mass, unless it has some means of shedding its angular momentum. No such means appeared forthcoming to the early supporters of the nebular hypothesis.

Another difficulty encountered by the theory is the lack of an obvious reason as to why a nebulous cloud should start rotating in the first place. Why should it not simply float through space like one of our familiar clouds drifting through Earth's atmosphere?

In contrast to the uniformism of the nebular hypothesis, another early suggestion postulated a catastrophic scenario in which the planetary system was formed from the splash of solar material created by the impact of a large object crashing into the youthful Sun. Getting this ejected material to condense into

FIG. 2.2 Pierre Laplace (Portrait by Jean-Baptists Paulin Guerin. *Courtesy*: Wikimedia)

planets did not prove to be an easy task however, but what if the impacting object (which had to be another star, despite the initial speculation that it might have been a comet) did not actually impact the Sun at all? What if, instead, there was a narrow miss?

In 1901, Sir James Jeans proposed and developed just such a scenario and in so doing offered the first serious challenge to the nebular hypothesis. In Jeans' scenario, a second star passed extremely close to the Sun, almost grazing the latter's photosphere. Tremendous tidal effects were generated on both objects. According to Jeans, the tidal surge on the Sun rose in a huge "mountain" of gas as the second star approached and then "darted out as a long tongue of gas" which stretched far into space as a long filament, teased out in the direction of the departing star. The gravitational pull of the second star was not, however, sufficient to carry this filament along with it and the tongue of material,

FIG. 2.3 Immanuel Kant (Portrait by unknown artist. *Courtesy*: Wikimedia)

though detached from the Sun, nevertheless remained within the latter's gravitational grip. The detailed theoretical investigation of the way in which such a filament would behave presents great mathematical difficulties, but Jeans conjectured that it might come to break up into a number of discrete condensations. The masses of these condensations are determined by the varying thickness of the original filament. Most of the filament's mass resided toward the mid sections, declining at either end. From this it follows that the more massive condensations will be the central ones, while those at each end of what we might picture as the cigar-shaped filament will be relatively small and light-weight. These condensations contracted to become the planets. As Jeans predicted, the truly massive ones—Jupiter and Saturn—inhabit the central regions while those at the sunward end were small by comparison, terminating in the diminutive Mercury at the very innermost tip of the filament. At the furthermost tip,

Uranus and Neptune, though large when compared to Mercury or Earth, are nevertheless lightweights next to Jupiter and Saturn. Although Neptune marked the edge of the known Solar System at the time Jeans proposed this model, the discovery of the tiny Pluto in 1930 seemed to fit his theory very neatly. In short, the model proposed by Jeans appeared to predict the structure of the Solar System very nicely and also avoided the question left unanswered by the nebula hypothesis, namely, what started the hypothetical cloud rotating in the first place. Moreover, unlike the nebula hypothesis, it did not necessarily have a problem with the slow rotation of the Sun. Consequently, Jeans' theory became very popular amongst astronomers during the first half of the Twentieth Century. It is a little ironic that a catastrophist theory should have gained such wide acceptance at a time when uniformism reigned in most scientific disciplines, but that is surely a testimony to the theory's apparent capacity to account for the observed features of the planetary system (Fig. 2.4).

FIG. 2.4 Sir James Jeans (*Courtesy*: uploaded to Wikimedia by Kokorik)

Nevertheless, the Jeans model had some difficulties of its own. On the face of it, this model provided no reason why the orbital plane of the planets should be close to the equatorial plane of the Sun. Nor does it explain why the planets should orbit the Sun in the same direction as the latter's rotation. On top of this, there is the reluctance to accept a chance catastrophic event as an explanation unless all else fails (as it appears to have failed for the genesis of the Moon, for example). Moreover, the feeling that planetary systems are somehow "natural" consequences of star formation was further strengthened by the discovery of proto-planetary discs around many young stars, much as would be expected if the nebula hypothesis was, if not necessarily exactly correct, at least on the right track. Then, by the end of last century, the discovery of planets around other stars surely clinched the case for these being regular side effects of star formation and not the products of some freak type of event.

Currently, the most widely-held hypothesis of the Solar System's formation is a version of the nebula hypothesis, albeit with some important differences. Today, the Solar System is thought to have formed through the collapse of a small region of one of the giant molecular clouds that are scattered throughout the Galaxy. Just as in the model put forward by the Eighteenth Century pioneers, this region of the cloud collapsed under the influence of gravity, with the bulk of the mass contracting into the central regions to become the Sun. Planets formed from the surrounding cloud through the accretion of solid grains, first into pebbles, then boulders, then larger mini-worlds known as planetesimals and finally into the planets we see today. Nevertheless, the Solar System is a different and more complex place than Laplace, or even Jeans, imagined or even could have imagined given the data available at these earlier times. Partly through the discovery of other Solar Systems—most of which are a lot less sedate than our own—and partly through the continued study of our own system at both observational and theoretical levels, we now believe that the Solar System has passed through eras of turbulence beyond anything that Laplace and his contemporaries could have contemplated. Even on the model of Jeans, the System remained a relatively quiet and sedate place after it formed from the filament left by that single, initial, catastrophe. Both Laplace and Jeans assumed that the planets formed pretty much where we see them today and both largely overlooked the presence of minor sub-planetary bodies as being of little consequence. The contemporary view of the Solar System is quite different. The early years were ones in which planets formed and migrated, some moving inward and others moving outward from their original places of birth. This migration was caused by their gravitational interaction with minor bodies; the remnant population of the vast numbers of planetesimals that did not snowball into major planets. Interaction between a planet and the myriad of planetesimals that it scattered every-which-way meant that it either gained or lost orbital momentum and consequently experienced either an expansion or a contraction of its orbit. Many of the planetesimals were either driven inward toward the Sun, flung out into interstellar space or pushed into large orbits far from the Sun. A good many of these objects were themselves large enough to be classified as dwarf planets and are still found to this very day orbiting in comparative obscurity out beyond Neptune. The strange little world Pluto—whose discovery appeared to boost the fortunes of Jeans' theory—is now known to be simply the closest, one of the largest and therefore (as seen from Earth) the brightest of these so-called Kuiper Belt objects.

Major planets also fell into orbital resonances with one another, meaning that relatively close approaches became established on a regular time scale, leading to drastic alteration of orbits over a longer lapse of time. Jupiter and Saturn were driven further apart and the "ice giants", Uranus and Neptune, relegated to large orbits far from the Sun. Previously, the presence of these two large planets so far out in space was seen by many as a problem for the nebula hypothesis. Simply put, there did not seem to be any way that sufficient material could be found so far from the Sun as to allow the formation of worlds as massive as Uranus and Neptune during the time taken by the formation of the Solar System. But given the "new" version of the hypothesis, with its mixture of uniformism and catastrophism (as we may fairly express it), this is no longer a problem. Uranus and Neptune actually formed much closer to the Sun where there *was* a sufficient supply of planet-forming material, and were subsequently tossed out into the surrounding wilderness. One version of the theory indicates that Uranus and Neptune even swapped places as the most remote major planet, although at that time neither was as remote as it is today. For all we know, there may have been other planets that got ejected from the Sun's family altogether and are currently wandering around somewhere in the depths of the Galaxy.

The problem of the slow-rotating Sun has also been solved by this amended version of the nebula hypothesis. Gas also accompanied the collapsing system, but the gas did not orbit the young Sun as quickly as the accreting planets. This difference resulted in drag which transferred angular momentum from the Sun to the planets, in effect putting the brakes on the Sun and pushing the planetary orbits further out into space.

A study published in 2011 suggested that Jupiter migrated inward until it was just 1.5 AU from the Sun. According to this same study, after Saturn formed it likewise migrated inward, though not as far, establishing a 2:3 mean motion resonance with Jupiter. That is to say, it completed two revolutions of its orbit for every three by Jupiter. This situation had the effect of pushing both planets outward again until they reached their respective positions where we find them today. However, according to this scenario, Jupiter's sojourn in a relatively small solar orbit had the effect of scattering planetesimals in the region where Mars now resides and it is argued that it is due to this culling of the planetesimal population in that region that Mars ended up having such stunted growth. This particular hypothesis does not appear to have convinced many workers in the field and critics argue that the prevailing conditions within the inner Solar System at the time when this was supposed to have taken place were not likely to have been conducive to the migration of the pair of giant planets outward to their present orbits. Nevertheless, whether this particular version is right or wrong, it aptly demonstrates the sort of chaotic processes that formed the planetary system that we know today and, although not everyone working in this field agrees on all the details and several versions of the theory exist, most conclude that the general idea is essentially correct. Nothing too weird remains … or does it?

But Is Gravity Enough?

Whether we look at the original nebula hypothesis, or its re-worked modern form, or even at Jeans' thesis, one common thread is seen to run through them all. The means by which the Solar System condenses from either a nebulous cloud or a filament ripped from the Sun, and the method by which the planets formed within that cloud or filament, is ultimately governed by a single force; gravity. On the face of it, there is nothing strange about this. Gravity is, after all, the dominant force on astronomical and cosmological scales. It may be nothing at atomic or molecular dimensions, but it reigns largely unchallenged when it comes to determining the structure and evolution of planetary systems.

Something that may easily be overlooked is the form that most of the visible matter in the universe takes. Unlike the material in our immediate terrestrial surroundings, much of the visible substance of the universe is not simply divided into the familiar states of solid, liquid and gas. On the contrary, the lion's share is in a fourth state; plasma. Not a great deal of this exists naturally on Earth, although we see it in such phenomena as ball lightning. But in the wider universe, the picture is very different. Plasma is defined in the Australian Pocket Oxford Dictionary as gas of positive ions and free electrons in about equal numbers. If we think about it for a moment, that describes just about everything from the stuff of stars to diffuse interstellar gas. In a very real sense, we live in a universe composed of plasma.

Well, so what? Plasma is still matter and it is still governed by gravitational force. True, but it is also far more susceptible to two other forces of nature—electricity and magnetism—than matter in the other three states. Is it possible that at least some of the tasks that we normally assume are carried out by gravity are really the work of electromagnetic forces governing the activity of plasma? There can be little doubt that electromagnetic plasma effects play some role, but a few scientists have developed the plasma scenario much further and worked out a model of the universe in which plasma not only dominates by its presence, but also determines the form taken by phenomena as diverse as galactic evolution and the formation of planetary systems.

Fɪɢ. **2.5** Hans Alfven in 1942 (*Courtesy*: Wikimedia)

One such scientist—indeed the principal proponent of the plasma model—was Hannes Alfven. Alfven was well suited to be the champion of this theory. It would not be an exaggeration to say that he was the world's leading expert in plasma physics. He actually invented the field of magnetohydrodynamics (mercifully referred to as MHD), for which he was awarded the 1970 Nobel Prize (Fig. 2.5).

A short critical account of Alfven's broader cosmological theory was already given a in *Weird Universe*, so a brief summary will suffice here to place his views on the genesis of the Solar System within their wider perspective.

Alfven's early ideas on the subject of plasma cosmology were published in his 1966 book *Worlds-Antiworlds*, and further developed in 1971 by his colleague O. Klein into what has come to be known as, either, the Alfven-Klein or the Klein-Alfven cosmology.

These authors begin by noting the puzzling observation that, although experiments using particle accelerators always produce particles and anti-particles in equal quantity, the universe at large contains very little anti-matter. This is usually explained by assuming that the creation process favored one form of matter or that a slight asymmetry in the amounts of matter and anti-matter exist (albeit too small to be detected in particle accelerator experiments) and, as Einstein remarked "matter won". The third way around the problem is the suggestion that matter was never created; that it has eternally existed. That may have seemed an option once, but it now flies in the face of just about everything we know of the universe and is not even worthy of serious consideration in the opinion of most contemporary cosmologists.

Alfven and Klein opted for yet another way out of the problem. They simply accepted what the particle accelerator observations revealed. Matter and anti-matter do indeed exist in equal quantities in the universe, but they are kept apart by cosmic electromagnetic fields formed between two thin boundary regions consisting of two parallel layers displaying opposite electric charges. Interaction between these boundary regions generates radiation, which in turn ionizes gas into plasma. Both matter and anti-matter particles are ionized in this way, resulting in a mixed plasma to which Alfven gave the name ambiplasma. The boundary layers are therefore formed of this strange matter/anti-matter substance. Although we might think that the matter and antimatter components of the ambiplasma should explosively annihilate one another upon contact, Alfven argues that because the particles and anti-particles composing it are too energetic and the overall density of the ambiplasma too low to allow rapid mutual annihilation, the ambiplasma should in reality be quite long-lasting. Moreover, the double layers will on the one hand act to repel clouds of particles of the opposite type while on the other hand attracting and combining clouds of particles of the same type. The upshot of this is a segregation of matter and anti-matter particles into ever-larger concentrations.

With a few minor exceptions, we only see matter simply because our region of the universe just happens to lie in one of the positive regions. But at great distances, beyond those accessible to our telescopes, this matter region gives way to an anti-matter region and because there is a slow annihilation of the matter/anti-matter particles along the boundary between these regions, the material within our visible universe slowly drifts outward. According to Alfven, it is this drift that we detect as the cosmological redshift of galaxies and mistake for the expansion of space. The Big Bang, Alfven opined, did not take place. Not at least, as the beginning of anything of fundamental cosmological significance. There may have been *a* big bang, but if there was, it was just one of the many violent events that take place from time to time in the universe. The universe itself, in the opinion of Alfven, had neither beginning nor end as far as can be ascertained.

As a cosmological hypothesis, this scenario encounters increasing difficulties as more details about the universe-at-large come to light. But Alfven does not confine the plasma hypothesis to this broad cosmological model. He sees evidence of plasma wherever he looks in the universe. The morphology of galaxies can be explained by understanding these as contracting balls of plasma. Much of what happens within the galaxies is also plasma related in his view; including the formation of Solar Systems like and, of course, including, our own.

Alfven voiced his dissatisfaction with what we might call traditional theories of Solar System formation in his Nobel lecture of December 11, 1970. He criticized the contemporary theories in so far as he saw them as essentially theories of the history of the Sun. Drawing comparison between the Solar System and the satellite systems of the giant planets, Alfven expressed the opinion that what should be developed is "a general theory of the formation of secondary bodies around a central body." The Solar System would then be seen as one application of this wider general theory. Not surprisingly, he saw this general theory as dealing with the evolution over time of a body of plasma.

The sequence of events through which all of the situations covered by the proposed general theory would pass can be summarized thus: "A primeval plasma was concentrated in certain regions around a central body and condensed to small solid grains [some of which may even have been present in the first place]. The grains accreted to what have been called embryos and by further accretion larger bodies were formed." Depending upon the specific application of this general scenario, these larger bodies may either be planets orbiting a star such as the Sun or moons orbiting large planets like Jupiter or Saturn.

Traditionally all of this is seen as the work of gravity, however Alfven reasoned that simply because the contracting mass is a plasma, other influences must come to bear upon the final outcome. In this, Alfven was resurrecting an idea that had previously been suggested by Arrhenius, whom we met in the previous chapter during our discussion of panspermia. What we actually have, according to Alfven and Arrhenius, is a mass of gas infalling toward a spinning central magnetized body while being ionized (i.e. turned into plasma) and brought into partial corotation with that central body. Grains condense from this plasma and, losing their electric charge, fall toward the equatorial plane where they are collected at various discrete distances from the central body. Sticking together, the grains first of all form streams of almost co-orbital particles to which Alfven gave the name of "jet streams". Within these jet streams, the grains continue to snowball together into larger and larger bodies. These are the planetesimals which, gently colliding and sticking together, form the final orbiting objects; be these planets orbiting a star or satellites in orbit around a large planet. Gravity becomes paramount only during the final stages of this accretion process.

Although this process, as applied to the formation of the major planets of the Solar System and principal satellites of the giant planets, was completed long ago, Alfven argued that minor examples of this form of accretion are found to this very day not too distant from our home planet. In a curious reversal of the accepted wisdom on such matters, Alfven and A. Mendis proposed, in the early 1970s, a scenario in which meteor streams become, as they expressed it, the source as well as the sink of comets and Apollo asteroids. In other words, they argued that the particles within meteoroid streams accrete together in the manner just described to form small planetesimals, completely or partially formed instances of which constitute the comets or asteroids generally designated as the parent bodies of such streams. Rather than being the parents of meteor streams via a process of gradual breakup and decay, these objects are seen by Alfven and Mendis as the children being born through the steady snowballing together of the meteoroids themselves.

Of course, both authors admit that there are instances where the existence of a meteor stream is clearly due to the breaking up of a parent body. The well-known instance of Comet Biela and the Andromedid meteor storms can hardly be explained any other way. This is, in other words, a clear case of the meteor shower being the sink.

Yet, referring to an earlier study by J. Trulson, Alfven and Mendis also argued that a stream of meteoroids might just as well act as the source of a small Solar System object. Following Trulson, they proposed that a meteor stream (understood as an example of a jet stream in the Alfven sense) that approaches Jupiter closely enough to be affected but not so closely as to be totally disrupted, can experience a velocity modulation of its constituent particles in a way that results in the creation of a travelling density wave. In itself, that will have no great and lasting consequence, however if the stream experiences two consecutive encounters with Jupiter, and two density waves are generated, it is possible for these two waves to interfere with one another in such a way as to create a region of significantly greater compression within the stream. Furthermore, if an appreciable amount of gas remains in the jet stream, we could no doubt add viscous effects into the mix as well and an even more enduring concentration could, in theory, occur. Denser concentrations of this type may condense still further on quite short timescales into something not too unlike a primitive planetesimal; either a comet or an Apollo asteroid. Alfven calculates that the entire process may complete itself on a timescale of just 90,000 years or thereabouts; a mere blink of the eye compared with the age of the Solar System.

Because they possess a rather large cross section, meteor streams are seen by these authors as a rather efficient means of mopping up interplanetary grains that were not originally members of the streams themselves. Because some of these grains, in regions of the Solar System more or less remote from the Sun, might be expected to contain water ice and frozen gases, some of the newly accreted bodies will contain volatiles and be capable of cometary activity while other grains originally orbiting at smaller distances from the Sun will be dry and the bodies forming from them will be inert asteroids.

The authors present as evidence favoring the recent formation of objects from meteor streams, the apparent late discovery of the so-called parent comets of two of the most famous of these streams; the Perseids and Leonids. Both of these showers have been known for centuries; references to them having been found in ancient Oriental records. Yet, according to Alfven and Mendis, their accompanying comets were only discovered during the nineteenth century. They remark that, if this conclusion is correct, the accretion of these objects from their associated meteor streams constitutes a "cosmic laboratory where we could still observe though on a much diminished scale the planetesimal process which led to the formation of the Solar System over 4.5 million (sic) years ago." (*Study of Comets* Part 2, p. 657).

Unfortunately for those wishing to see a scaled down version of this process, the late discovery of the two comets cited by these authors does not mean that they only formed during the last couple of centuries. The Leonid comet (55P/Tempel-Tuttle) is an intrinsically rather faint one which only makes naked-eye visibility during a very close approach to Earth. Close approached did, however, take place in 1699 and earlier in 1366 and on each occasion a fast-moving comet was observed. Orbital computations have identified each of these with Tempel-Tuttle. The Perseid comet (109P/Swift-Tuttle) is, by contrast, intrinsically rather bright, but is frequently badly placed and, having a period of around 130 years, is not a frequent visitor to our skies. Nevertheless, it has now been identified on a small number of occasions in ancient records; the first of these being over 2000 years ago.

The Alfven-Mendis model has not won many followers, to put it mildly. Despite the theoretical possibility of concentrations in meteor streams arising from the interference of density waves, the process seems an unlikely one to occur in nature with the required intensity to produce the results that the theory demands. Meteor streams are very diffuse and the prospect of such snowballing seems rather remote. The masses of some of the associated objects would also demand that their streams contained a great deal of material in their youth. Comet Swift-Tuttle, for example, appears to be very large with estimates of its nucleus as high as 30 km (18.6 miles) in diameter. A great number of meteoroids are also required to build a decent sized Apollo asteroid. Moreover, the requirement of gas being present in the streams has no observational justification and seems unlikely at best. On the other hand, the process of steady disintegration of comets and some asteroids has both observational and theoretical bases and we have clear evidence (as already mentioned and as admitted by Alfven and Mendis) that at least some meteor streams arise from this process. Without strong evidence to the contrary, there seems no compelling reason not to believe that all of them have a similar genesis.

Planets Spun from the Sun?

Although the plasma theory of Alfven and colleagues differs in certain respects from what has become the widely accepted form of the nebular hypothesis, it is still broadly within the bounds of that scenario. Like the orthodox model, it sees the planetary system as accreting from a cloud of material surrounding the infant Sun.

Nevertheless, despite the theoretical and observational support for this thesis, a few dissenters remain. One of these was the very controversial Thomas Van Flandern to whose hypothesis we shall now turn (Fig. 2.6).

Basically, Van Flandern objected to the dependence that all forms of the nebular hypothesis placed upon particles colliding and sticking. He writes "At first, the sticking results in pairs of molecules here and there. Eventually, some of these collide with each other and stick, and bodies of larger and larger size ultimately are build up … the literature supposes that high velocity collisions are destructive, whereas low-velocity collisions are 'constructive' i.e., result in sticking" (*Dark Matter Missing Planets & New Comets*, p. 328). He admits that small bodies that collide with larger ones

FIG. 2.6 Thomas Van Flandern (*Courtesy*: American Astronomical Society)

will tend to accrete there, but is mystified why gently-colliding objects of approximately equal mass should "stick" together. He doubted that the relative velocity of the particles would be low enough and wondered where the supposed "stickiness" was supposed to originate. Once larger bodies are involved, gravitational attraction takes over the accretion process, but the initial problem (as Van Flandern saw it) is how the process is able to reach that rather late "gravitational" stage at all.

Nevertheless, even this later stage of accretion encounters problems in explaining the growth of planets in nearly circular orbits around the Sun as well as moons in similar orbits around their primary planets. Accretion through gravity works but, Van Flandern argues, only if the orbits of the accreting bodies are mildly chaotic. In the case of circular orbits, objects sharing the orbit of a larger body are forced to librate back and forth and thereby avoid collision with the larger mass.

Van Flandern also raises the old problem of the slow-rotating Sun as well as pointing to the difficulty of planets as distant as Neptune accreting from what must have been very diffuse material so far from the central hub of the Solar System. Both of these apparent difficulties now look a lot less daunting since the time of Van Flandern's writing, thanks to the further development of the nebula hypothesis.

Finally, Van Flandern points to the similarity in chemical composition of Solar System objects as evidence that they formed under a smaller range of temperatures than the nebula hypothesis required. Specifically, he noted the presence of abundant carbon dioxide on the surface of Neptune's moon Triton, arguing that this is very difficult to explain if this body had formed in situ and had in consequence never been much warmer than it is at present. Once again however, more recent models of the early Solar System allow for a good deal of mixing of materials, so that it is not surprising to find high-temperature and low-temperature species together in the same object, as evidenced by analysis of dust collected by the *STARDUST* probe from the coma of Comet Wild 2. Also, the acceptance of the phenomenon of migration of planets means that we are no longer tied to the assumption (one could rightly say presumption) that Solar System objects have always been where we see them today. Quite the contrary, in fact.

Van Flandern does not claim to present a fully worked out Solar System model of his own, but he does provide an outline of the form which he believes that such a model would assume.

Broadly speaking, he apparently agrees that the early Sun formed from a contracting cloud of matter. Through the conservation of angular momentum, the proto-Sun must, he argues, "reach an overspin condition numerous times during its contraction phase." (DM, p. 331). Each time this happened, a blob of matter would be thrown off, taking with it some of the proto-Sun's angular momentum, and thereby causing its spin to decrease to below the critical velocity. Yet, because the proto-Sun continues to contract throughout this process, it will not be long before conservation of angular momentum again spins it up to critical velocity and the process begins all over again; another blob is ejected and some more angular momentum is lost. Eventually, the proto-Sun settles into its stable state and is no longer "proto". What eventuates now is a young star surrounded by blobs of matter thrown off at points where its rotational velocity had become critical. Tidal forces would, Van Flandern argues, act on these orbiting blobs in such a way as to enlarge their orbits. Moreover, the blobs of material would start out with the forward spin of the proto-Sun and must therefore all be rotating in the same direction, something which the nebular hypothesis does not readily explain in Van Flandern's opinion. Moreover, each would go through the same sort of contraction and fission process experienced by the proto-Sun, albeit on a much reduced scale. In time, the blobs of material thrown off by the contracting proto-Sun would become the major planets, complete with their systems of satellites.

Van Flandern admits that the notion of planets being spun off the Sun had previously been presented and rejected on the grounds that the present rate of rotation of the Sun is well below the critical level, plus the observation that the Sun's equator does not quite lie in the plane of the planetary orbits, being just 7° off the mean plane of the planetary system. These objections do not worry him however. He suggests that both conditions can easily arise if the interior of our star spins much faster than its surface and if the tilt of the Sun's spin axis was altered during the contraction phase by interactions between the faster-spinning core and the slower-spinning outer regions. Van Flandern considered that the evidence available at the time of his writing (concerning the differential rotation rates between the solar surface at the equator and at the poles) pointed to a rapidly-rotating solar interior. Later research has not, however, confirmed this. The outer regions of the Sun rotate fastest at the equator and slowest at the poles and this pattern continues throughout the convective zone within the Sun's interior. However, the situation changes at the boundary between the convective and radiative zones, known as the Tachocline. This boundary is situated at one third of the solar radius. Rotation changes, at the Tachocline, to a solid-body form and continues in this form throughout the radiative zone, right to the Sun's center. The rate of this solid-body rotation is about equal to that of the surface at middle latitudes—intermediate between the fast equatorial and the slow polar values.

In view of these findings, the rotation of the Sun remains a serious problem for Van Flandern's hypothesis. Moreover, as mentioned earlier, more recent discoveries of pre-planetary nebula surrounding other stars and the evidence that planets are accreting within some of these in the manner postulated by the nebula hypothesis seems to supply direct observational evidence that this process actually works in nature. Planetary migration adds a further complicating feature. While this process is included in Van Flandern's model to a certain degree, the outward expansion of orbits such as he envisions is very mild fare when compared with what is now thought to take place during the early ages of planetary systems. Our own system must have been truly chaotic, yet it pales in comparison to what must have taken place in some (possibly most) other planetary systems in the universe. But that is a good thing for us. Had the Solar System gone through what most of the newly discovered systems experienced, we most probably would not be here at all.

Exploding Planets

For the present however, we will overlook the difficulties inherent in Van Flandern's general model of the Solar System and take a look at some specifics. This is where we find that Van Flandern's Solar System was not only formed in a manner differing from the more widely held models, but also had some physical differences from the system that any of these other models predicted.

The early Solar System, according to Van Flandern consisted of Venus and its moon (which eventually broke away to become the planet Mercury), Earth, Planet V, Planet K, Jupiter, Saturn, Uranus, Neptune and their various moons; one of Neptune's eventually escaping to become Pluto. Our own moon and what is now the planet Mercury each formed from their respective parent planets by fission and what we now call Mars formed in a similar manner from either Planet V (orbiting roughly where Mars is today) or Planet K which supposedly was a superterrestrial of between 5 and 20 Earth masses that orbited at a solar distance of around 2.8 AU; roughly in the middle of today's asteroid belt, close to the orbit now occupied by the dwarf planet Ceres.

Now what, we may ask, became of the planets designated as V and K? This is where things get truly weird—according to Van Flandern, both of these worlds exploded.

The idea of an exploded, or at least shattered, planet goes back to H. Olbers in the early Nineteenth Century. This was the time when the first asteroids were discovered and Olbers put forward the quite reasonable-sounding hypothesis that such objects were the remnants of a planet that once orbited between Mars and Jupiter but that had for some reason broken apart in a truly catastrophic event. The idea went out of fashion in favor of the less catastrophic idea that the small bodies within that region were essentially partially formed planetesimals that were prevented from coming together to form a true planet because of the disrupting effect of the gravitational tug of giant Jupiter. Vestiges of the Olbers approach were resuscitated however as the explanation for groups or families of asteroids sharing very similar orbits around the Sun; the so-called Hirayama families, named in honor of Kiyotsugu Hirayama who discovered them in 1918. Most of these groupings are believed to have arisen from the breaking up of larger parent asteroids, most probably following collision with other bodies within the main asteroid belt, although groupings of smaller objects might also arise from the rotational instability of rapidly spinning asteroids.

Nevertheless, in the 1970s, Canadian astronomer M. Ovenden revived the original broken planet model. He harked back to the

1814 development of Olbers' hypothesis by J.L. Lagrange which supposed that both asteroids and comets were the remnants of the hypothetical disintegrated planet. According to Ovenden, the planet not merely disintegrated but exploded with such violence that its debris was scattered all over the Solar System; some the rocky material going into sedate orbits in the asteroid belt while much of the outer layers got hurled to the farthest reaches; much of it leaving the Solar System altogether. The fragments that did not quite go careering off into interstellar space are returning today as long-period comets. By the way, the name given by Ovenden to his hypothetical exploding planet was Krypton, not implying that it was filled with the inert gas of that name, but in honor of the mythical home planet of the comic-book character Superman and which, according to the story, suffered a similar fate. This is why Van Flandern calls his second exploding planet "Planet K"—second because he also argued that Planet V blew up at an earlier date in the Solar System's lifetime. The demise of Planet K was, however, a very recent event in terms of Solar System history. According to Van Flandern, it happened only around three million years ago.

The main support for the exploded planet hypothesis, in Van Flandern's opinion, is the existence of the asteroid belt itself. He argues that bodies even as small as the average asteroid could not have accreted in their contemporary position if the Solar System of ancient times had been much the same as we see it today. Each exploded planet contributed to the asteroid population of the Solar System. The innermost of the two, Planet V, was named as the parent of S-type asteroids; dry, rocky bodies from which ordinary chondritic meteorites are almost certainly derive. Planet K gave us C-Type asteroids, the most common objects in the outer regions of the asteroid belt and the ones believed to have very similar composition to carbonaceous meteorites.

Van Flandern also sees evidence of the explosion in the nature of the orbits of asteroids and comets. Looking first at the orbits of asteroids within the so-called main belt between Mars and Jupiter, Van Flandern admits that some of the original explosion signatures of these would have disappeared. For instance, although all the orbits must originally have passed through the point in space that the planet occupied at the time it exploded, precession would have long ago eliminated this common point.

Nevertheless, several patterns would continue to be present. For example, there would be a "minimum eccentricity ... for every mean distance from the Sun" (*DM*, p. 186). Orbits at the mean solar distance of the point of explosion would be circular, while those with smaller or greater distance "must have at least enough eccentricity at some time that they can reach the mean distance of the explosion." Such evidence does, he claims, exist. Although some of the S-type asteroids in the inner regions of the main belt do not conform to the prediction (a discrepancy which he suggests might be explained by these objects being collisionally evolved) there is, he argues, a trend found within the orbital eccentricity of main-belt asteroids such that "the further their mean solar distance is from 2.8 AU, the greater is the minimum eccentricity of all the orbits found at that mean distance" (*DM*, p. 186).

But orbital evidence favoring a planetary explosion is not, he argues, confined to the asteroid belt. Indeed, if the explosion occurred about three million years ago and if it ejected fragments almost to the escape velocity of the Solar System, these fragments should now be returning to roughly the point of their explosive birth on elliptical orbits having periods of around three million years, give or take a fair amount of scatter due to stellar perturbations and the like. Dynamical studies of first-time or dynamically new comets do indeed indicate periods of this order and Van Flandern sees this as evidence supporting his exploding planet thesis. However, an object coming into the central planetary system from a distance close to the maximum for something still gravitationally bound to the Sun will naturally have periods of this order irrespective of whether that object began life as a piece of planetary shrapnel shot out at just under escape velocity three million years ago or whether it experienced gravitational acceleration to near escape velocity billions of years ago and has ever since been parked in a precarious orbit about as far from the Sun as it could go without drifting off into the interstellar void. The latter scenario developed from the one put forward by J. Oort and which is widely accepted today. As it stands, the appeal to the periods of these dynamically new objects alone does not decide whether Oort or Van Flandern is correct.

It is only fair to remark however, that Van Flandern's concept of the Oort Cloud understands its supporters as hypothesizing that this great sphere of distant objects condensed in situ from the outermost fringes of the contracting pre-solar nebula. That seems to have been the opinion of E. Opik, who came to the distant comet reservoir conclusion independently of (and actually prior to) Oort. At least, he mentioned this genesis of the Oort Cloud as a possibility although he did not work it out in detail. Van Flandern is rightly critical of this version of the hypothesis, pointing out that at the distances being discussed, matter would have been so thinly distributed that accretion into solid bodies of any size could not have taken place. Nevertheless, Oort was in complete agreement with Van Flandern as to the impossibility of comets forming at the distances he postulated. He saw the formation of these objects as a natural consequence of planetary accretion and one that took place relatively close to the infant Sun, in the same region where the planets themselves formed. In his scenario, the accretion of comets represented a stage in planetary formation. Most of the comets snowballed into the planets themselves, with much of the remainder being gravitationally expelled from the Solar System. Those eventually taking residence in the Oort Cloud were the ones that escaped assimilation into planets but were ejected at velocities not quite reaching that of the escape velocity of the Solar System. This picture is, broadly speaking, the one most widely accepted today. In so far as it depends upon ejection from the region of the planets, certain similarities may be drawn between it and Van Flandern's model, while noting that the ejection took place during the period of planetary formation over four billion years ago and, even then, was not a consequence of a planetary explosion.

Van Flandern also draws attention to another aspect of the dynamics of these dynamically new comets which he claims supports his hypothesis. He reminds his readers that "If comets did originate in a breakup event in the inner Solar System, then the original distance of closest approach to the Sun ... for their orbits must all have been less than or equal to the distance at which the breakup occurred." (DM, p. 189). From this, he predicts that "there will be fewer comets with perihelion points inside the Earth's orbit than for a comparable range outside the earth's orbit" and finds that this is indeed reflected in their orbital statistics. Nevertheless, to prove the prediction made in the first of these quotes, those same orbital statistics should also reveal a marked decline in the numbers of comets coming to perihelion at distances greater than about 2.8 AU, where Planet K was supposed to have been located. Early statistics may indeed have appeared to support this, but only because comets at large distances will inevitably be faint objects, except for the very few of unusually large size and high intrinsic brightness. Thanks to the deep searches of the past couple of decades however, comets with perihelia out past 9 AU are being picked up and we can be sure that even now most of the fainter ones are passing by unseen. There is clearly no falling away of numbers once 2.8 AU is passed. Furthermore, the number passing within Earth's orbit does not appear especially small for this volume of space and some have been observed to come very close to the Sun itself. The closest approach on record for a comet that can definitely be identified as having come straight in from the Oort Cloud was made by Comet ISON in late 2013. This rather small object approached the Sun to just 0.01 AU and was reduced to a cloud of dust for its impertinence.

Van Flandern's theory of comets does not stop with his account of their origin. Not surprisingly, his model of the physical nature of these objects also differs in some important respects from the one widely held today, but more about this in Chapter Four of this book.

Following the planetary explosion, according to Van Flandern, a cloud of dark carbonaceous material spread throughout the Solar System and dusted the planets to a greater or lesser degree. Objects having a slow rotation were most conspicuously affected by this cosmic dust storm. One body which Van Flandern sees as having been especially marked by this event was Saturn's moon Iapetus. This is the moon with one relatively bright and one unusually dark hemisphere and it is easy to understand why such a discrepancy of surface albedo should have been especially interesting to Van Flandern. Nevertheless, the explanation for the two-faced character of this object appears more straightforward than he opined. The presence of Saturn's dark-surfaced moon Phoebe is generally thought to be the culprit. Dark carbonaceous dust particles ejected from the surface of this moon appear quite capable of explaining this odd characteristic of Iapetus. Phoebe orbits Saturn beyond Iapetus and travels in a retrograde direction. Particles ejected from its surface likewise follow its retrograde path and, as Iapetus pursues its prograde orbit around the planet, these are swept up by its leading hemisphere. Nevertheless, the full explanation of the dark hemisphere of this moon does not stop there. Because the material from Phoebe is so dark, it absorbs more energy from the distant Sun than the surrounding, icy, surface of Iapetus, causing the dark region to become warmer than the surrounding areas. In the warmer micro-environment, volatile substances evaporate, leaving behind dark material native to Iapetus itself and darkening the region still further. This in turn increases the absorption of solar energy and so on, with the upshot being the evaporation of ices from the dark hemisphere and their recondensation on the bright one, leading to the extreme contrast that we see today.

Evidence for this model has been observed in the form of a ring of dark dust particles in the orbit of Phoebe, presumably composed of particles that have recently been ejected from the surface of this moon. The reason for this ejection of particles is widely thought to be due to the impact of meteoroids, however there is another possibility which the present writer has raised from time to time. Almost certainly, Phoebe is not a native moon of Saturn. It is thought to have once been a free-orbiting body that was captured by the giant planet's gravity, as evidenced by its retrograde orbit and large distance from Saturn itself. In all probability, it was once a centaur similar to Chiron. These objects appear to be large comets and activity has actually been observed on some of them, including Chiron itself. It does not seem unreasonable to think that Phoebe may have experienced periods of cometary activity, maybe triggered by a meteorite strike cracking away an area of insulating surface crust and exposing a fresh reservoir of frozen gases. Maybe the dark ring and the black face of Iapetus were caused by dust ejected during one or more such episodes. Nevertheless, whether raised by meteoroid impacts or by cometary activity on Phoebe, it is clear that the dust coating the leading hemisphere of Iapetus can be explained more simply than by involving the presence of an exploding planet.

Such difficulties with the model as those cited above pale in comparison with the deeper problems that it raises. In the opinion of the present writer, the type of recent catastrophic event envisioned by Van Flandern has two overriding difficulties. First, why should a planet, formed billions of years earlier, suddenly explode and, secondly, what effects would such an extreme event have on the nearby worlds in the Solar System?

Looking at these in reverse order, it seems to me that something as dramatic as Van Flandern postulated would have greater effect than a few impact craters and a dark coating on one side of a Saturnian moon. The sort of event he pictured is demonstrated by the opening paragraphs of his chapter on the topic of planetary explosions in *Dark Matter*. He begins with an account of the eruption of Nova Cygni 1975! Unfortunately, his treatment of the nature of this event could be confusing for today's reader as the process he describes is really that of a core-collapse supernova and his remark that the Sun will end this way in the very distant future is no longer believed to be the case. This just means that the interpretation of novae has moved on in recent years and it is now recognized that these are not single stars that for some reason blow up but are members of an especially violent category of what are known as cataclysmic variables. These objects involve a pair of stars in close orbit around a common center of gravity. Both stars are very old and have departed from the hydrogen-burning main sequence but one of them (presumably the more massive of the pair during its main sequence sojourn) has already gone through the stage where its core has collapsed into an ultra dense degenerate state—a white dwarf as these degenerate cores are called—and the outer layers have been puffed off into a planetary nebula. This, it is now believed, is what will happen to our Sun billions of years from now. The other member of the pair has only aged to the red giant stage. That is to say, it has swelled out enormously into a distended and relatively cool envelope. Stars pass through this stage prior to the white dwarf/planetary nebula phase and smaller stars take more time to reach it than larger ones (very small and very large ones have different destinies entirely, but that is another story). Cataclysmic variables occur when the white dwarf pulls material away from the distended red giant. This material spirals down onto the white dwarf and collects on its surface, building up until a critical mass is reached and the layer of accreted material goes off like a thermonuclear bomb on steroids! Neither star is any the worse for the explosion. Indeed, it acts as a safety valve for the white dwarf. Sometimes, for reasons not yet clear, the explosion does not occur as described above and so much material is collected that the entire white dwarf is converted into an exploding bomb; an inconceivably violent event known technically as a Type 1 supernova. In the milder classical novae however, the glowing gases in the explosion's aftermath may outshine the combined light of both stars by anything from a few hundred to a million times for a relatively brief period of time. Nevertheless, the system soon settles down and the accretion process begins all over again, ready for the next time.

This picture of a nova, although given briefly and oversimplified here, is pretty much established through detailed observations. They are not really exploding stars and no nova ever observed hints at having been an exploding planet; which is just what Van Flandern suggested. In other words, according to his hypothesis, if an alien civilization in a neighboring planetary system happened to be looking skyward around three million years ago, they would have seen a nova near a yellow star which we now call the Sun. They would have seen something similar to what we earthlings witnessed in 1975.

Think about this for a moment. Early hominids walked the Earth in those days. What effect would a second Sun, far and away outshining the first, have had on them? What effect would the blast of radiation have had on life in general? While there is no detailed assessment of the effects of a nova within the Solar System, but it seems intuitively obvious that surface life on Earth would have been wiped out and, most probably, the planet's atmosphere and oceans dispersed into space as well.

Then there is the vexed question as to why a planet should explode in the first place. Nothing we know about the internal constitution of planets leads us to suppose that they are prone to blowing up; which is good news in view of the fact that we live on one of them. A definitive answer is not given by Van Flandern, although he does make several suggestions, some marginally more plausible than others. On page 163 of *Dark Matter*, he writes "we have only recently learned that thermonuclear processes similar to those in a star may now be taking place in the core of the planet Jupiter, which emits more heat than it takes in from the Sun." A similar remark was once made by the equally controversial V.A. Firsoff, who speculated that deuterium fusion might be occurring within the core of Jupiter. It is true that deuterium fusion does take place in objects that are similar to Jupiter in composition, however we now know that the mass of even that giant planet is insufficient for this reaction to have started up in its core.

The heat that this giant planet emits is a product of its steady contraction. Giant Jupiter is getting smaller, albeit by only about 2 cm (not quite 1 in.) every year. Still, this contraction is enough to heat the core of the planet to something like 36,000°; hot indeed, yet still too cool to trigger thermonuclear reactions.

In order to reach the temperatures and pressures required for deuterium fusion to begin, Jupiter would need to be some thirteen times more massive than it is. It would then be classified, not as a planet but as the smallest and most lightweight form of brown dwarf; one of those transition objects that fill the gap between massive planets and lightweight true stars. Increase the mass of Jupiter to 65 times its present value and lithium would also start fusing in its core, transferring into the class of more massive brown dwarfs. A further increase to 75 times its real mass would enable hydrogen to fuse and it would finally graduate into the class of true stars, albeit only to become one of the smallest of the faintest members of the category of so-called red dwarfs. Yet, even if (counterfactually) conditions within the core of Jupiter did allow atomic fusion to take place, there is no reason to expect that it would explode. We don't see brown dwarfs or red dwarfs exploding. In fact, we have no reason whatsoever to believe that such events occur. Furthermore, Jupiter did not explode. Van Flandern's hypothetical exploding planets were much less massive than Jupiter, so what happens—or fails to happen—within the core of this planet has no relevance to the supposed demise of Planets V and K.

But if fusion reactions cannot perform the required task of blowing a planet to smithereens, perhaps fission can work. Van Flandern raises the speculation that natural fission reactors—concentrations of radioactive materials deep within planets—might sometimes reach critical mass and, in effect, turn into naturally occurring atomic bombs on a truly gargantuan scale. Considering the comparatively rare nature of sufficiently radioactive materials, this seems rather farfetched to put it mildly, although the question of possible natural reactors or even atomic explosions on a small scale has been raised by other researchers in association with some geological features. Van Flandern points to evidence of natural chain reactions in Gabon. Such reactions cannot happen today because of the depletion over time of U-235, but around 1800 million years ago there was four times the amount of this isotope and chain reactions were possible. The chance that these could blow up an entire planet, however, seems pretty farfetched and that suggestion would require a lot more evidence before it is taken seriously. Moreover, the planetary explosion in which Van Flandern is most interested is supposed to have taken place just three million years ago, not nearly 2000 million.

Van Flandern's next speculation comes very much from the proverbial left field. He points to the discovery of high energy gamma rays from the galactic center and notes that the energy of these (511,000 eV; an enormous energy in comparison with the feeble 2 eV of visible light, as Van Flandern is quick to point out!) is just what would be expected to issue from the annihilation of 10^{10} tons of positrons (anti-electrons) every second. He further notes that gamma rays having the fantastic energy of 1.8×10^6 eV have also been detected coming from the galactic center and sees this as evidence that several stellar masses of the short-lives radioactive element Aluminum 26 is continually decaying there. The production of this element is usually associated with supernova explosions, but Van Flandern suggests an association with the positron annihilation taking place at about the same location. He then states that "Al 26 is also found in carbonaceous meteorites" (*DM*, p. 164), which in his opinion represent part of the debris from the exploded planet, and from this he draws the tentative conclusion that "the suggestion is clear that antimatter might have played a role in the explosion of the missing planet too. Although clearly no detailed mechanism can yet be proposed, the same lack of a detailed mechanism is true about the galactic center". From this tenuous position, he then makes a further leap that maybe gamma-ray bursts arise from "Antimatter from the planetary explosion's blast wave encountering interstellar matter" (*DM*, p. 164). The mind boggles at the prospect of such an antimatter blast wave travelling outward through the Sun's planetary system. Planets, asteroids, comets and interplanetary particles would all become little gamma-ray burstars as the antimatter wave front engulfed them and mutual matter/antimatter annihilations took place at their surfaces and (in the case of the planets) within their atmospheres. If such a thing had really happened just three million years ago, it is very difficult to imagine that terrestrial life could have survived. It is far from sure that any remnant of Earth itself would have survived.

The claim that AL 26 is found in carbonaceous meteorites should also be clarified. The half-life of this isotope is only 7.17×10^5 years, so its presence in meteorites could not even date from the time of the supposed planetary explosion. This isotope is created in meteorites by cosmic rays and is useful in determining the length of time that a meteorite "find" has been residing on Earth. Once the parcel of matter that becomes a meteorite is exposed to space, either through the breakup of its parent body or by exposure on the surface of a parent object, it is constantly bombarded by cosmic rays and it is this process that produces Al 26. After the meteorite falls to Earth, the atmosphere shields it against cosmic radiation and the production of this isotope ceases as, all the while, the Al 26 already within the meteorite decays at a rapid rate. The terrestrial age of the meteorite can then be determined by comparing the amount of Al 26 remaining. This is a useful tool provided by nature, but as it stands it has little to do with exploding planets.

The most exotic, and by far the weirdest, hypothesis concerning the cause of the explosion relies on an unconventional theory of gravity which Van Flandern adopted from G-L Le Sage who first postulated it in 1748, and N.F. De Duillier who proposed something very similar, even earlier in 1690. In this hypothesis, gravity results from a flux of invisible ultra-mundane corpuscles to which Van Flandern gives the name of C-gravitons. These particles are thought, by their few supporters, to impinge upon all objects from every direction at velocities exceeding that of light itself. In a paper given by Van Flandern in 1998, he asserted that they travel at a velocity at least twenty billion times greater than photons of light. He also thought that they could provide limitless free energy and even put them forward as a means of providing superluminal space travel. But amidst all these hopeful properties, there was a negative. All masses in the universe are inevitably absorbing energy from the influx of these particles and if the source of energy which they provide is truly limitless, what would happen to a planet if something stopped it from reaching thermodynamic equilibrium by ridding itself of the energy that it was receiving? He speculated that such might happen if an insulating layer was formed within a planet due to a change in state in that planet's core. Unable to rid itself of this limitless influx of energy, the planet would simply blow itself to pieces (Figs. 2.7 and 2.8).

The biggest problem with this hypothesis is the very dubious state of the hypothetical C-gravitons. Nearly everyone who has

FIG. 2.7 Georges-Louis Le Sage (*Courtesy*: Wikimedia)

FIG. 2.8 Nicolas Fatio de Duillier (*Courtesy*: Wikimedia)

worked on the theory of gravity has rejected the Le Sage model as lacking in evidence. Gravity remains a mystery, but Le Sage does not appear to have provided the solution and, if this theory fails, another mechanism proposed for exploding planets fails together with it.

Captured Moons and Martian Theme-Parks

As already mentioned, Van Flandern did not believe that Mars was numbered amongst the Sun's original planetary family. Not, at least, as a true planet. It was a moon of one of the worlds that later exploded, probably Planet V which seems to have been the more earthlike of the two and which orbited close to where Mars is found today. As a consequence, Mars was not initially accompanied by its two tiny moons, Phobos and Deimos. In a rare display of harmony with the orthodoxy prevailing at the time of his writing, Van Flandern agreed that these Martian moons are captured asteroids. Indeed, he even went so far as to state that they are "certainly former asteroids" (DM, p. 272). This is no longer as certain as it may have appeared to be at the time he wrote those words, however. In fact, it now seems that the moons resulted from a giant impact during the youth of the Solar System, rather like a scaled down version of the impact that is widely thought to have led to the formation of our own satellite. In part, this change of opinion has come about through the difficulties encountered in getting captured asteroids into the orbits followed by Deimos and Phobos. Although there might be a certain intuitive appeal about the capture hypothesis, getting it to work in detail has proven to be far from satisfactory.

Van Flandern was very aware of the difficulties involved in the orthodox version of the capture hypothesis and it was here where he dramatically parted company with the more conventional supporters of that theory.

For one thing, because he saw asteroids as recent additions to the Sun's family, he did not accept the prevailing thesis that the capture of the moons took place long ago in the life history of the Solar System. Indeed, as Mars itself was not a free-orbiting planet until Planet V blew up, it must have been moonless for most of its existence. All this changed about three million years ago in Van Flandern's opinion when Planet K exploded. By that time, Mars was orbiting the Sun. Many months after the explosion, the smaller pieces of planetary shrapnel reached Mars, but their velocity was too high for the small planet to trap them into orbit and most of them simply passed by. Some presumably impacted the surface but a few passed through the Martian atmosphere and, for a substantial range of altitudes, would have been slowed sufficiently by atmospheric friction to go into temporary orbits around the planet. Many of these orbits would have decayed after only a very few revolutions due to the continuing atmospheric drag. Nevertheless, the total number of these temporary moons may have been considerable and by the time the slower-moving, larger, fragments of Planet K arrived, a screen of the tiny temporary satellites surrounded the planet. The larger masses collided with these and "Among the many such collisions likely to occur, some may be suitable to leave a few of the larger masses gravitationally bound to Mars in a permanent way" (*DM*, p. 274). Continuing collision with the smaller and very temporary satellites drove these larger captured bodies into more circular and somewhat more equatorial orbits. Several small moons may have formed in this way, although Deimos and Phobos remain the only pair that managed to survive down to the present day.

Van Flandern sees the decaying orbit of Phobos as evidence for a recent capture. This moon also has remarkable grooves on its surface, which Van Flandern sees as evidence that this moon had at one time run into streams material in Martian orbit, just as expected in his scenario. Moreover, both moons resemble C-Type asteroids; just the sort of material that he believed the exploding Planet K spread throughout the Solar System.

Probably, the biggest hurdle to accepting this account of the genesis of the Martian moons is the requirement of an exploding planet in recent Solar System history. But when we turn from the moons of Mars to some of the surface features of that planet, Van Flandern's ideas become distinctly weirder. He speculates that, prior to its explosive demise, Planet V was inhabited, not merely by animal life, but also by civilization. Upon seeing the signs of the coming catastrophe, these beings migrated to their largest moon (the future planet Mars) and survived there until the second planetary explosion. It seems that they noticed in advance that Planet K was in trouble and had enough time to set off for the apparent safety of Earth before it blew itself to pieces. Van Flandern suggests, albeit only as a hypothesis, that these "intelligent inhabitants of the dying planet might have intervened directly in the evolution of species on Earth, so as to produce an intelligent species here" (DM, p. 341). In that way, the strange coincidence between Planet K's explosion and the appearance of early hominids on Earth ceases to be a coincidence at all. Alternatively, he suggests that the extra cosmic radiation from the blast of Planet K may have sped up terrestrial evolution or, maybe, killed off some natural enemy of the primates, thereby allowing this species to multiply and diversify.

Now, if a civilization had once been present on Mars, we might expect that remnant signs of this might still be observable. In the opinion of some radical thinkers, this is exactly what has been found. These are the famous—or infamous depending upon your point of view—pyramids and face. The face in particular has drawn a great deal of wild speculation, although most of this was based upon early and relatively low-resolution images. Later images show the superficially artifactual appearance of this formation to dissolve into random markings. In Dark Matter, Van Flandern takes a skeptical view of the subject, in the true sense of the word "skeptical". By that, I mean that he leaves open for further investigation both the possibility that these features found on the surface of Mars are artifacts of a vanished civilization and the possibility that they are simply natural formations which superficially resemble something artificial. He apparently leaned toward the second more conservative opinion stating that "Most probably [they are] natural rock formations which by chance had taken on forms of interest to humans, the way a cloud in the sky might occasionally do if one is looking for shapes (DM, p. 346)."

It is indeed interesting to note just how often faces and human forms in general appear in natural formations. In the present writer's home state of New South Wales, there is a hill overlooking the tiny village of Wingen in the Hunter River Valley that bears a striking resemblance to the profile of a woman. The Wingen Maid is even named on maps of the area, yet nobody has suggested that it is artificial. Moving to a more unearthly example, can you detect the "smiley face from outer space" in the image of Stephan's Quintet (Fig. 2.7)? Van Flandern's statement in *Dark Matter* is surely correct. And yet, he has over time become associated with the more extreme opinion that these formations really are the work of an ancient civilization. In later articles and public lectures he even suggested that the face might have originally been constructed as a sort of theme park or museum. In Australia, we have structures like the Big Banana, Big Marino and so forth which seem to resemble the sort of thing about which he was thinking. Did Mars perhaps have a Big Face? (Fig. 2.9).

One wonders if Van Flandern really, in his heart of hearts, believed any of this or whether he was simply stirring the pot to get people to think outside the box of conventional scientific method. He apparently believed in the possibility that this could be correct, but that is not the same thing as believing in the actuality. What his real thoughts were can only be surmised.

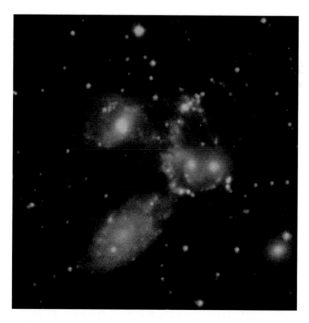

Fɪɢ. **2.9** Stephan's Quintet, complete with "smiley face" (*Courtesy:* NASA)

Project 2.1: The Faces in Outer Space

Simply as a matter of curiosity, examine some of the images of the surfaces of Mars, Pluto, Mercury and so forth taken from flyby or orbiting spacecraft and see how many faces and other formations can be found that might be cited as artifacts by someone committed to finding evidence of intelligent extraterrestrial life. I am not suggesting that any of these formations really *are* artifacts, but it is interesting to find just how often artificial patterns can be read into perfectly natural formations. Try it and see for yourself.

Be that as it may, it is difficult to believe that any hypothetical "Planet V-ians" could have survived all of this turmoil. Granted that Planet V must have been a smaller planet than Planet K and so is unlikely to have gone off with the force of a classical nova explosion, it still seems incredible that an intelligent race could have simply hopped over to one of the planet's moons and then used this as their successful Noah's Ark while the parent planet blew up all around them. Although we don't know what triggered the explosion, it must have been violent and, if it also marked the birth of S-Type asteroids in the manner that Van Flandern speculates, an awful lot of planetary debris must have whizzed past the moon Mars, with a great deal not going passed at all but actually impacting the surface. As we earlier saw, Van Flandern gives several possible mechanisms by which a planet might explode, some of which would have resulted in a rather nasty blast of radiation. None of these looks very convincing, as we argued, but if the planet actually exploded and was not simply struck by a huge meteorite (which must have been rare in those days, unless an even earlier planet exploded within the inner Solar System) one might conclude that some violent process producing a good deal of radiation was probably involved. Given that Planet V was, according to this scenario, apparently inhabited by creatures with sufficient intelligence to build spaceships capable of transporting them to one of its moons but not necessarily intelligent enough to avoid building atomic weaponry, one might even dare to suggest that they blew the planet up themselves. Van Flandern does not consider this possibility however, and there are some clear objections to it. His hypothesis of exploding planets is clearly based upon the occurrence of natural processes. Moreover, if the inhabitants of Planet V destroyed their own world, it is probable that they would not have known of the catastrophe in advance and would not have had time to prepare for, and execute, their escape, though one might argue that they already had a colony established on Mars and that it was the inhabitants of this that survived. In any case, the survival of life on the moon of an exploding planet remains a very dubious scenario.

Nevertheless, a worse situation must eventually have followed with Planet K going up in an even bigger blast. Apparently Mars, long since a planet in its own right by that time, was deemed too close for comfort, but I strongly doubt that Earth would have been in a much safer position, as mentioned earlier. Maybe bacillus infernus, living deep down within the planet's crust, might have survived, but I doubt if anything else would have, be that native or alien life.

Alien Intervention in Terrestrial Biology?

Going back a little way to the earlier remark regarding the possible effect on terrestrial life of these hypothesized planetary explosions, we recall that Van Flandern suggested the possibility of a very direct influence via the manipulation of terrestrial life by intelligent beings from Planet V (via Mars) who fled to Earth when they saw Planet K giving signs of its impending blast. Overlooking the difficulties raised by exploding planets in general and their devastating effect upon life on other planets within their home systems in particular, we might wonder whether this recurring theme of alien intervention rests on anything that might be construed as firm evidence. Alien intervention does not depend upon the hypothesis of an exploding planet. It does not even depend upon Mars or one of the other Solar System worlds having been inhabited by intelligent space-faring beings at some time in the past. If we are going to speculate, let's be really daring and suggest that somewhere in the Galaxy a race long ago achieved interstellar travel and visited Earth three, four, or more millions of years ago. At least, let us suggest this if only for the sake of speculation. Is there anything that might allow this suggestion to be taken seriously?

Several decades ago, the name Erich von Daniken made headlines around the globe following the publication of his best-selling book *Chariots of the Gods?* and its sequels. Very briefly, what this author suggested was that ancient beliefs and legends of gods and heroic beings might actually have been based on half forgotten accounts, handed down from one generation to the next, of actual contacts between human beings and visitors from another planet. The "gods", in short, might have been alien astronauts!

This book created quite a stir, to put it mildly. Although *Chariots* was presented as a possible explanation for stories of gods and so forth (as was implied by the question mark at the end of its title), many folk simply accepted it without question as fact. Book sellers at the time remarked that many customers were purchasing the more expensive hard cover editions rather than the cheaper paperback and interpreted this as indicating the value that they were placing on the work. It was something to be kept and given a place of prominence amongst their collection of books, not a quick-read paperback that might then be discarded or donated to the local charity book stall.

In the subsequent books, von Daniken assumed a more dogmatic and extreme position and his ideas came under increasing examination by skeptics. His claim that ancient artifacts were probably the work of extraterrestrials was criticized by many anthropologists who argued that far simpler explanations, involving only humans and their hard work and capacity for innovation, were quite adequate to account for the statues, temples and rock artwork of ancient times. Some of von Daniken's supposed evidence was exposed as downright fraudulent. For example, a ground marking shown in a photograph in one of his books which he claimed to be the remains of a prehistoric airport was found to be a tiny feature hardly large enough to fit a model aircraft. It looked superficially convincing, as long as there was nothing to demonstrate its scale. Add a human being, towering over the marking, and its true dimensions became all too apparent. Allegations of even more extreme misrepresentations have been made. For example, one critic claimed that the discovery of certain alleged ancient works of cave art apparently revealing incontrovertible evidence of alien visitation and shown to von Daniken by a native of the region in which they were found, apparently had no existence outside the author's mind. The person who allegedly revealed this ancient treasure to von Daniken was later questioned about the subject, and informed his interviewers that they had the story back the front. He had not told von Daniken about anything. It was von Daniken who had informed *him* about the ancient art.

Thanks to this increased skepticism about that author's claims, von Danikan's "Chariots" have by now largely ridden off into the sunset taking their extraterrestrial "gods" with them.

Von Daniken is mentioned only twice by Van Flandern in *Dark Matter*, but it is clear that the latter took his views seriously in the sense that he considered them to be worthy of proper scientific investigation. He even went so far as to state that he believed the hypothesis of contact between humans and extraterrestrial visitors was superior, in explaining the similarities found in ancient artifacts and stories, to the competing orthodox theses. He argued that the ancient astronaut hypothesis was more obedient to the scientific principle of Occam's Razor in so far as it involved fewer assumed hypotheses than the more widely accepted opinions. In view of the criticism of von Daniken however, that position is difficult to justify. Even if we are very charitable toward von Daniken and excuse his falsification of some of the presented evidence as due to his unbridled enthusiasm in getting his point across, it remains true that no artifact put forward as being possibly alien was clearly and demonstrably outside of the manufacturing capability of the humans alive at the time it was constructed. We must rid ourselves of the idea that people of our own time and culture are unique in their abilities to invent clever ways of doing things. A quick glance at the Roman aqueducts, the early Buddhist temples and mediaeval cathedrals is enough to ensure us that there were great buildings before what we like to call the modern age. And if we admit these as human works, then why not extend the same courtesy to, for example, the ancient Egyptians and their sphinxes and pyramids?

The latter, although often featuring in lists of wonders that some people believe required superhuman intervention, can be clearly seen as human works if we trace their evolutionary development from simpler forms of tombs. Egyptologists found that these earlier types of tombs were rectangular structures with flat, slab-like roofs and outward sloping sides. Such a structure is known as a mastaba, literally meaning "house for eternity". The next phase was effectively placing one of these structures, or something closely approximating it, on top of another and thereby growing a step pyramid. These were the earliest forms of Egyptian . pyramids, prior to the final phase which we might call the filling in of the steps to give the smooth and regular . pyramids which became so symbolic of that country. While Egyptologists might be gnashing their teeth at such a brief account of. pyramid evolution, the general picture is sufficient for our present purpose. What it shows is that there is no need to introduce extraterrestrials at any point in the process (Fig. 2.10a and b).

What really would turn the tables in favor of the hypothesis of alien intervention would be something like the discovery of an ancient . pyramid of pure titanium, a transistor radio in a 2000-year-old undisturbed tomb or a silicon chip in ancient geological strata. Thus far, no such artifact has turned up.

Maybe von Daniken was too controversial even for Van Flandern, as his mention of this author is brief, as already noted. More space, albeit mostly in an extensive footnote, is afforded in his book to Zecharia Sitchin. Unlike von Daniken, Sitchin did not amass examples of ancient artifacts which he claimed could not have been constructed without alien help. Rather, he relied on ancient writings, claiming that stories of encounters between humans and gods or other supernatural beings had their origin in real encounters with extraterrestrial visitors.

Briefly summarized, Sitchin's hypothesis states that at the beginning of human existence, extraterrestrials landed on Earth and genetically engineered a species of hominid (humans) to serve their labor requirements. These genetically engineered hominids were made in the likeness of the extraterrestrials themselves and commanded to serve their masters. They were also kept in ignorance of certain subjects, knowledge of which their extraterrestrial masters considered to "dangerous" for them to acquire. None the less, the inquisitive humans managed to acquire this knowledge anyway, and were cast out of the presence of their alien masters and driven away from their protection. However, the humans continued to multiply and even cross-bred with the aliens themselves. The situation became so bad that the aliens decided to wipe them from the face of the planet, although not before they selected a very small number of superior humans whom they protected from the cleansing catastrophe. Nevertheless, subsequent

FIG. 2.10 (a, b) The evolution of Egyptian tombs from mastaba to stepped pyramid (*Courtesy*: Egypt archives)

descendents of these chosen humans sought to imitate the aliens by constructing a spaceship capable of carrying them to the dwelling place of the alien masters themselves. This attempt was especially worrying to the aliens and they became deeply concerned that humans might go on to develop even more advanced technology; with possible dire consequences. In order to put a stop to this human arrogance, the aliens caused a scattering of the human race through the confusion of their languages. Apparently, even this did not work and the aliens eventually gave up on the human race and went back to their home planet.

If this sounds vaguely familiar, that is no surprise. It is essentially a re-writing of the Book of Genesis with the assumption that God referred to the alien master race. Further speculation by Sitchin has Sodom and Gomorrah being destroyed by nuclear weapons and the Pyramids of Egypt constructed as nuclear bomb shelters.

Van Flandern laments that most of his colleagues irrationally react negatively to such considerations as these. As an example of what a more rational reaction might be, he suggests that "sediment samples from the bottom of the Dead Sea should be quite definitive in determining whether or not nuclear weapons were used on Earth long ago" (DM, p. 340 f). The reader might like to construct a mental fantasy about a researcher approaching his or her department for funding to conduct a sampling of the Dead Sea floor to determine if nuclear weapons destroyed the cities of Sodom and Gomorrah back in the days of Abraham! Somehow, I think that funds would be directed elsewhere, but not because of closed-mindedness amongst scientists, as Van Flandern believes. It is simply that for a suggestion to be transformed into a true hypothesis, hard evidence must be forthcoming. Yet, the evidence for Sitchin's scenario is weak to say the least. For one thing, if a highly technological alien race had lived on Earth for thousands of years, one might expect some remnants of their technology to have survived, at least in ruined form. This point was made earlier in the context of supposed ancient alien artifacts, but it is relevant here as well. It is one thing to say that the pyramids were nuclear bomb shelters and that "this makes ever so much more intuitive sense than the concept of a tomb for a pharaoh" (DM, p. 340f), but when no evidence of nuclear blasts in the Egyptian desert (forget the Dead Sea for the present) has been found, when mummies of pharaohs turn up in pyramids, when the evolution of pyramids from earlier tombs is quite obvious and when the entire tomb building enterprise so clearly blends with the ancient Egyptian view of death and the afterlife, the bomb-shelter idea not only ceases to make more intuitive sense, it begins to verge on the absurd. It might, in our own culture, appear more fruitful to spend the immense investment of money, time and labor on nuclear bomb shelters than on tombs, but that does not fit with the thinking of an ancient Egyptian. Even if nuclear weapons had existed in those distant times, there is no certainty that shelters would have won the day over tombs. After all, what would have seemed more important to an ancient Egyptian; preservation of life in this world for a few more years or keeping it for eternity?

What is arguably the biggest problem encountered by all of these hypotheses of ancient astronauts or whatever we might call them is that nothing has turned up in archaeological digs that is radically inconsistent with the technological achievements of human culture as it existed at the estimated age of the material found within these ancient sites. This alone suggests, as already stated, that in the absence of very strong evidence to the contrary, humanity was responsible for whatever artifacts are found. At the risk of laboring the point, we repeat that if a laser is ever dug up in a Stone Age tomb or a fossilized computer chip found in million year old sediment, then the prospect of ancient alien intervention will at least have some empirical evidence that is convincingly solid enough for serious consideration. But drawings on cave walls which might represent spaceships if seen with the eyes of faith but might just as easily be diving birds with their wings folded are simply not enough to support such farfetched speculations.

Nevertheless, whilst we make no apologies for being critical of Van Flandern's hypotheses, we welcome any ideas that make us think again about the hypotheses that tend to get taken for granted and treated almost as absolute truth. We remember that in the late nineteenth century, one of the eminent scientists of the day wrote that the most firmly established scientific fact of all was the existence of the ether, the supposed all-pervading substance believed necessary for the transmission of electromagnetic waves. Not long after these words were spoken, Michelson and Morley, in attempting to measure the ether drag as Earth passed through this substance, actually succeeded in demonstrating its non-existence. Then, closer to our own time, the British Astronomer Royal made his (in)famous statement to the effect that space travel "is all bilge" just a few years prior to the launch of the first artificial Earth satellite and just over a decade before astronauts landed on the Moon. Scientific theories and opinions can become dogma if not challenged. We need our Van Flanderns with their unconventional and downright weird theories. Whether we agree with them or not, they provide the challenges that open our minds to other possibilities. May they keep coming!

3 Focused Starlight, Cosmic Impacts and Life on Earth

There was a time, not so very long ago, when the idea that our planet is from time to time engulfed in some megacatastrophe capable of altering the course of history and even of the evolution of life itself, would have been given short shift in scientific circles. Once the true age of geological formations became determined and Darwin's theory of a slow development of terrestrial life gained wide acceptance, it seemed to many people both inside and outside of the scientific community that nature proceeded at a very slow and steady rate, and to suggest otherwise amounted to little more than superstition. That is no longer so today, but prior to this relatively recent shift in opinion, any catastrophist hypothesis was seen as weird or naive. Still, a few brave souls swam against the tide and in doing so paved the way for the broader understanding of natural processes that we hold today.

Nature's Ray Gun?

One of the weirdest and most ingenious of these catastrophist hypotheses was spawned, albeit indirectly, by Einstein's theory of General Relativity. Consideration of one of the consequences of this theory led to a suggestion which however, was dismissed, even by Einstein himself, as being too far into the left field to be given serious attention. That turned out to be a big mistake; up to a point at least.

The aspect of General Relativity of relevance here is the prediction that light itself bends in the presence of a gravitational field. This is not in itself an innovation of Relativity, as even Newton's

© Springer International Publishing Switzerland 2016
D. Seargent, *Weird Astronomical Theories of the Solar System and Beyond*, Astronomers' Universe, DOI 10.1007/978-3-319-25295-7_3

theory of gravity predicts that there will be some degree of deflection as a light ray passes close to a massive object. Nevertheless, Einstein's thesis that what we call the force of gravity can be better interpreted in geometric terms as the warping of the very fabric of the space-time continuum itself in the vicinity of a massive body, yields a significantly greater bending of light than a prediction based upon Newton's theory.

As is well known, the first test of this prediction arrived with a total eclipse of the Sun in 1919. Fortunately, the position of the Sun at the time of the eclipse lay near some rather bright stars. This meant that the light from these stars must pass very close to the Sun's limb and it was realized that if a photograph could be taken of these stars during the progress of the eclipse, their precise position could then be measured and compared with their measured position on photographs taken at night, when the Sun was nowhere near that region of the sky. If the light from these stars was bent by the gravitational field of the eclipsed Sun, these measurements should differ slightly. Furthermore, the degree to which they differed would show whether Newton or Einstein was right. As just about every introductory book on relativity records, the prize went to Einstein.

The result was welcomed as strong evidence in favor of General Relativity, but beyond this no consequences of the phenomenon were considered. Apparently, however, Einstein had thought of one back in 1912, even before he had published his theory. Einstein figured that if light can be bent by the gravitational field of a massive object, it should also be possible for a strong field to focus light, somewhat in the manner of a glass lens. In this instance however, gravity (that is to say, warped space-time itself) acts as the lens. In theory, a distant object should be magnified if its light passes through the strong gravitation field of a foreground body. Rather strangely, Einstein paid no further attention to this idea and seems to have essentially forgotten about it.

Although Einstein himself later called his postulation of a cosmological constant "my biggest mistake", we might more accurately use this expression for his dismissal of the phenomenon of gravitational lensing as the process is known today. The phenomenon was first observed in relation to quasars. Back in 1979, what appeared to be two identical . quasars flanking a galaxy of far lower red shift were shown to be two images of the same . quasar gravitationally lensed by the mass of the intervening galaxy. For pointlike sources such as . quasars, four separate images are actually formed flanking the intervening massive object, although only in the best resolved instances can all four images actually be observed. The best known example of this is the so-called Einstein cross in the constellation of Pegasus. The cross consists of four images, having magnitudes between 17 and 19, of a single quasar gravitationally lensed by an intervening galaxy located some 400 million light years away from Earth (Fig. 3.1).

A remarkable quadruple image of a very distant supernova, gravitationally lensed by a galaxy within the cluster MACS J1149.6+2223 was discovered on Hubble images of 2014 November 10 by Patrick Kelly of UC Berkeley. This "Einstein cross supernova" provides a further remarkable demonstration of gravitational lensing (Fig. 3.2).

FIG. 3.1 The Einstein Cross (*Courtesy*: ESA/Hubble & NASA)

Fɪɢ. **3.2** Einstein cross supernova (*Courtesy*: Hubble Space Telescope)

Project 3.1: Finding the Einstein Cross

Can the Einstein cross be observed visually in amateur telescopes?

The answer is yes, but with some reservations. This is a task for experienced deep-sky enthusiasts who have access to telescopes with apertures that are large by amateur standards. Still, thanks to John Dobson, remarkably large reflectors are now more readily available to amateurs and more telescopes capable of locating the Einstein cross are found in astronomy club observatories and even in the possession of private observers than in pre-Dobsonian days.

The object may be found at

RA = 22 h 40 min 30.3 s
Dec. = +3° 21′ 31″

The lensing galaxy, PGC 69457, is around magnitude 14 and as such should not pose too many difficulties for telescopes of 16-in. (41-cm) diameter or thereabouts. The lensed image of

(continued)

(continued)

the quasar has been glimpsed as a blurred star near the edge of the brighter central regions of the galaxy in telescopes of 18 in. (46-cm), but resolving the four images of the cross requires considerably larger apertures. Experienced observers have, however, reported catching fleeting glimpses of all four images while using telescopes of 24-in. (61-cm) aperture.

Smaller scale instances of the phenomenon (termed gravitational microlensing) have been used since 1986 to detect planets orbiting stars within our own galaxy. That year, Princeton astronomer Bohdan Paczynski drew attention to the fact that individual stars should also act as gravitational lenses focusing the light of any background stars that they may eclipse. Because of the weaker gravity of the intervening mass (a star rather than an entire galaxy) the multiple images would remain too close together to distinguish with current technology (they would be separated by less than 0.001 seconds of arc), but the brightening of the "eclipsed" star should still be apparent and, if that star had any sufficiently massive planets orbiting it, secondary brightening caused by these should also be apparent under the right circumstances. Several planets have now been found using this method.

Distant galaxies may also be gravitationally lensed as their light passes through intervening foreground clusters of galaxies. In these instances, the distant galaxy images are distorted into long and narrow arcs. Nevertheless, the true shape and dimensions of these objects can be deduced from these distorted images, allowing information to be gleaned about galaxies that, without the benefit of gravitational lenses, would remain out of reach of contemporary telescopes; galaxies that are truly far, far away. In a very real sense, intervening clusters of galaxies act as nature's own telescopes, revealing objects that even our most sensitive instruments could not otherwise observe.

As readers of *Weird Universe* will recall, the person who really brought this phenomenon to light was not Einstein. It was Rudi Mandl , a professional engineer and amateur physicist. Early in the 1930s, Mandl wrote several letters to Einstein expressing his opinion that the gravitational lensing effect could not merely be detected but may have played a significant role in the evolution of life on Earth. Einstein, in part put off by this latter suggestion (about which more will be said shortly) seemed to consider Mandl to be verging on the crackpot. Frustrated by Einstein's lack of attention to his ideas, Mandl eventually travelled to Princeton to confront the famous physicist in person. In response to Mandl's tenacity, Einstein yielded and published a short paper in 1936. In a subsequent letter to *Science* however, Einstein made it clear that he considered the phenomenon to be of no practical or observational importance and that he only formally brought it to the attention of the wider scientific community because "Mister Mandl squeezed [it] out of me" and that its publication "makes the poor guy happy".

In *Weird Universe* it was argued that Mandl's contribution to the recognition of this important phenomenon has gone largely unrecognized for too long and it was proposed that the double arc-like images of gravitationally lensed remote galaxies be semi-officially referred to as Mandl arcs in his honor. It is to be hoped that this suggestion is acted upon, as this will at least bring his name into wider acknowledgment and hopefully lead to a more general recognition of the role which he played in astronomical history.

In extreme cases, the two Mandl arc images join up into a horseshoe shape or even a complete circle, known as an Einstein ring. An impressive example of this phenomenon is the image of a very remote galaxy distorted into an almost-complete ring by the gravitational lensing of the intervening galaxy LRG3-757 as shown in Fig. 3.3.

Whilst not detracting from his insight, in all fairness we should also point out that in one respect Mandl went too far. At the other end of the spectrum from Einstein's dismissal of the phenomenon as nothing more than a useless curiosity, Mandl saw it as playing an important role in the history of Earth's biology. This was, as Einstein correctly assessed, verging on the crackpot and yet, even here Mandl reached a conclusion that was ahead of his time.

In short, what Mandl proposed was that the gravitational lens effect might, under certain circumstances, focus stellar radiation onto Earth with such intensity as to trigger genetic mutations amongst terrestrial organisms. In place of the steady evolutionary track of orthodox Darwinism, Mandl saw changes in terrestrial life as (at least in part) arising in jumps as species mutated and as some of these genetic mutations acquired characteristics that gave them greater survival potential than the earlier species.

FIG. 3.3 Einstein Ring (*Courtesy*: ESA/Hubble & NASA)

This is a load far too heavy for gravitational lensing to carry. It is simply not strong enough to have the sort of ray-gun effect as suggested by Mandl. Yet, even though his proposed mechanism for altering the course of our planet's biological history cannot do the job that he proposed, his hypothesis of genetic jumps rather than slow and steady Darwinian evolution and his suggestion that at least some of these jumps may have an extraterrestrial origin has become more acceptable in recent decades than during the period in which he gave voice to these speculations. He may have been wrong with respect to the mechanism, but he was surely correct about the occurrence of dramatic episodes punctuating the course of terrestrial life. He also seems to have been partially right in suggesting that at least some of these episodes had their origin beyond the confines of our planet.

Biological Extinctions, Asteroidal Impacts and Other Things

About 66 million years ago, a dramatic change occurred on Planet Earth. The Cretaceous era—an age during which dinosaurs still roamed the Earth—came to what in geological terms could be called an "abrupt" end, giving way to the Tertiary era during which mammals rose to the dominance that the great saurians had formerly enjoyed. Clearly, some major climatic and ecological upheaval must have taken place resulting in mass extinctions and the emergence of a new ecosystem. Although this event is popularly most noted for the extinction of the dinosaurs, many other forms of life also vanished then. Indeed, it is estimated that about 17 % of all families, 50 % of genera and 75 % of all species became extinct at that time.

The cause of this sudden catastrophe has long been debated with suggested answers ranging from an extreme outbreak of volcanism or some other purely terrestrial event to astronomical causes such as the impact of a large asteroid or a supernova explosion dangerously close to the Solar System.

. An important clue was unearthed in the late 1970s when Luis and Walter Alvarez of the University of California at Berkeley discovered an abnormally high concentration of iridium in the thin layer of clay marking the boundary between the Cretaceous and Tertiary eras (the so-called K-T boundary or, as it is increasingly known, the K-Pg or Cretaceous-Paleogene boundary). Levels of iridium within this layer reached six parts per billion. Although this may not seem to be high, compared to an average for the Earth's crust of just 0.4 parts per billion it is a very significant concentration. The most striking feature of this high iridium concentration however, is that it constituted strong evidence that an impact of a very large meteorite must have occurred at the time during which the boundary sediments were formed. Meteorites are known to have far larger iridium concentrations than terrestrial rocks—up to around 470 parts per billion—so the relatively high concentrations in the K-T layer would seem to be best explained by the impact of a large meteorite followed by the diffusion of meteoritic dust around the globe during the following years. Mixed with ordinary terrestrial dust, this iridium-rich material became diluted and settled into the discovered clay layer, giving rise to iridium levels which, while less than that of the impacting body, were nevertheless elevated above normal terrestrial concentrations (Fig. 3.4).

At the time of this discovery, no crater was known dating from the time of the K-T boundary. A number of suggestions were, however, proposed. One interesting speculation put forward by the well-known and respected astronomer Fred Whipple suggested that the impact may have been so great that the impacting object

FIG. 3.4 K-T boundary layer exposed by erosion on a rocky hillside, Badlands near Drumheller, Alberta (*Courtesy*: G. Larson)

might actually have penetrated Earth's crust. If that is really what happened, the resulting scar would not be a crater. On the contrary, it would be an up welling of fresh magma, probably creating a new area of land. Whipple suggested that we should look for a relatively young and strongly volcanic island, pointing toward Iceland as a good possibility. This suggestion, though interesting and refreshingly outside the box nevertheless fell by the wayside for lack of supporting evidence (Fig. 3.5).

More promising as a candidate was the Chicxulub crater, an astrobleme or ancient impact crater buried beneath the Yucatan Peninsula in Mexico. The first major hint of something interesting in this area came as far back as 1951 when exploratory borings in search of oil hit what at the time was thought to be a dome of solidified lava some 4200 feet (1.3 km) beneath the surface. This was unusual, as lava domes are not a feature of the area. No further odd features were found, however, until the late 1970s when a geophysicist named Glen Penfield, at the time employed by the Mexican state-owned oil company Pemex, discovered another

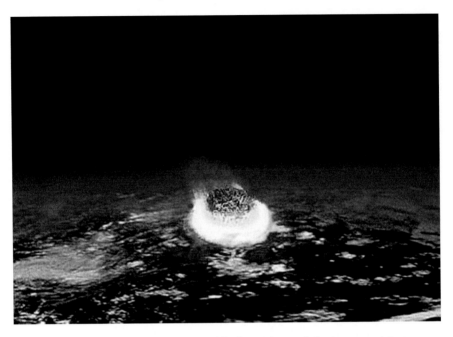

FIG. **3.5** An artist's impression of the Chicxulub impact (*Courtesy*: NASA)

strange geological feature buried beneath the peninsula. This looked like a buried crater, but Penfield was unable to confirm its exact nature at the time. Nevertheless, through contact with Alan Hildebrand of the University of Arizona (who had been informed in 1990 of Penfield's discovery and of the suspicions as to its nature raised by journalist Carlos Byars), he later obtained samples supporting the impact-crater interpretation of the feature. The presence of shocked quartz and tektites in surrounding areas, together with detection of a gravitational anomaly, strongly implied that the feature was indeed the site of an ancient impact by something very large indeed.

Later in the decade of the 1990s, examination of satellite images revealed the true size of the ancient crater. It is huge, having a diameter some 300 km (190 miles) across! The blast needed to gouge out a hole of that size is estimated to have been of the order of one hundred million megatons or around two million times more powerful than the largest thermonuclear device ever exploded by mankind. To trigger a blast of this magnitude, the impacting object is estimated to have been around 10 km (6 miles) in diameter. Such a body would have been quite capable of blowing enough meteoric dust into the air to account for the iridium-rich layer discovered by Luis and Walter Alvarez (Fig. 3.6).

The effects of such a massive blast as this would have been devastating. It seems that the region where the asteroid struck was covered by deep water at that time and under such circumstances we can imagine that enormous tsunamis—veritable mountains of water towering thousands of feet into the air—radiated outward from the point of impact and, upon making landfall, sweeping away everything in their path. Moreover, the impact must have generated great clouds of super-heated steam, as well as thick

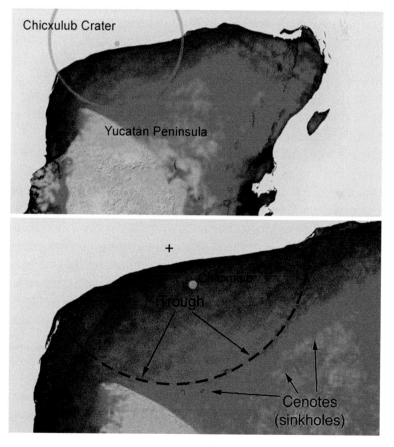

FIG. 3.6 Images of the Yucutan Peninsular showing the Chicxulub Crater outline (*Courtesy*: NASA)

billows of dust and ash, that boiled up into the atmosphere in a stupendous mushroom cloud, dwarfing any of those fearful formations unleashed during the atomic age. If that was not frightening enough, shock waves racing outward from the site of impact through the surrounding terrestrial crust violently shook the ground and may have triggered secondary earthquakes and even volcanic eruptions as the planet reacted to this cosmic attack with plutonic violence of its own. An outburst of volcanism, triggered by the cosmic impact, would have injected sulfate aerosols into the atmosphere, adding still further particulate matter to the already large quantity ejected skyward from the blast itself. Sulfate aerosols had two major effects on the environment, neither of which would have been confined to the immediate area of the impact and neither of which would have been good. Aerosols ejected high into the atmosphere, high enough to reach the stratosphere, would in time come to enshroud the entire planet in a high-altitude haze, reducing the light and heat received from the Sun and leading to a period of global cooling. Aerosols that failed to reach these altitudes, on the other hand, would wash down in rain, significantly adding to the acidity of precipitation and causing injury to much of the land's vegetation.

But even before these detrimental effects became widespread, great wildfires undoubtedly swept large regions of the planet. Huge volumes of smoke and ash, together with carbon dioxide cooked out of the carbonate rocks in the impact zone may have spawned, first of all a period of global warming due to the greenhouse effect, followed by a severe cooling as the matter released from the fires joined the other high-altitude aerosols, adding further to the Sun screen blanketing the Earth. It is thought that so much particulate matter would have been wafted into the upper atmosphere that a significant darkening of the daylight hours may have lasted for several decades. Photosynthesis was interrupted, leading to global die-back of many plant species. This directly affected the survival prospects of herbivores and indirectly that of the carnivores preying upon them. This reduction in plant life may have been partially offset by the increased carbon dioxide in the atmosphere, but the trauma to the entire ecosystem inevitably resulted in numerous extinctions, especially of species which may for one reason or another have already been struggling to survive.

Popular discussions have given the impression that this was the dinosaur's downfall; the whole story of the cause of their extinction. This is sometimes followed with a chilling reminder that a similar event, should one occur today, would almost certainly see the end of humanity.

Nevertheless, many paleontologists remain unconvinced that this scenario tells the whole story of the dinosaur demise. These creatures seem to have been on the decline even before the K-T catastrophe occurred. The reasons for this are probably quite complex. Some scientists have argued that the large dinosaurs had simply grown too big to be as efficient as a changing world demanded. That might indeed be true, but not all dinosaurs were giants and the small ones disappeared together with their larger relatives. Or did the small ones really disappear after all? Some argue that they are still with us today, except that we now know them as birds.

This latter thesis has also become more complex in recent years with the discovery of fossilized bird skeletons dating back millions of years prior to the K-T boundary. Apparently birds not unlike those of the present time co-existed with the dinosaurs, in the later stages of the latter's reign at least. Some of these ancient birds failed to break through the K-T boundary, but others quite clearly did. For reasons that are not at all clear, birds which could broadly be called duck-like survived the catastrophe. These were not strictly speaking ducks in the modern sense, but were broadly similar in their general characteristics.

Furthermore, paleontologist Robert Bakker points to the fact that frogs were also contemporary with the dinosaurs and even depended upon a similar environment. If the effects of the Chicxulub impact were as deadly to all species as traditionally portrayed, how did frogs manage to cross the K-T boundary while dinosaurs did not?

It seems that some species were more sensitive to the climatic disaster caused by the Chicxulub impact than others, but it is not really obvious why some survived while others became extinct. Size alone was apparently not the sole deciding factor as both the largest organisms (the large dinosaurs) and some of the smallest (plankton) suffered severely, as did some middle-sized organisms (birds that were not like ducks, for instance) while others seemed not to have been too badly affected. Seen in this light, it rather looks as though the K-T extinction might not have been entirely due to the cosmic impact. Maybe the impact was simply the proverbial last straw.

To pursue this would take us too far from our topic however. For our purpose, it is sufficient to say that the extinction of the dinosaurs' and the wider topic of the effects of the Chicxulub impact appears to be more complex than it is often presented as being and that the last word on the subject has almost certainly not been written.

What is no longer in doubt is the occurrence of an enormous collision about 66 million years ago and the massive climato- logical and ecological shock that resulted from this. Whether the meteorite was the total cause or not, its arrival saw the disap- pearance of one era of life on this planet and the birth of another. Yet, however serious and devastating the event was, life survived and the ecosystem restored itself in what was, geologically speak- ing, but a moment of time. If the thought of an asteroid collision engenders fear, this corresponding fact should equally inspire hope in the long-term survival of our blue world.

What was the Asteroid's Origin?

The size of the body that struck Earth 66 million years ago was greater than any asteroid capable of making close approaches to our world today. Moreover, it would appear that the Earth was hit by a number of smaller objects (in the order of tens of meters across) around the same time as the Chicxulub meteorite. At least two smaller craters are known from that era (Silverpit in the North Sea and Boltysh in Ukraine) with a third (Shiva) of disputed impact origin. Then, going back further into Solar System history, the Moon was struck by an asteroid somewhat smaller than the Chicxulub object around 108 million years ago, giving rise to the conspicuous Tycho crater.

It seems that the last 60 million years have been pretty quiet in the Earth-Moon neighborhood compared with the previous period of equal duration. It would appear that there was a height- ened influx of asteroidal bodies into the inner planetary system at that earlier time. Was this spike in asteroid impacts simply one of those clusters of events that appear in random distributions, or did it result from a common cause? And if the latter, what might than common cause have been?

In any hunt for a common cause, it will be of help to determine what kind of object actually struck the Earth at that distant date. In general terms, there are really only two broad alternatives. Either it was an asteroid or it was a comet. Considering that more asteroids than comets have been discovered in orbits that are potentially hazardous to Earth, the impacting object was most probably an asteroid. Nevertheless, some astronomers argue for the comet alternative on the ground that the object was unusually large for the class of asteroid that we find in Earth-crossing orbits today. That is a valid point, however by and large the asteroid alternative seems the more likely one.

But what kind of asteroid could it have been? Was it rocky or metallic, and, if the former, was it typical of an inner main belt S-type body or was it a C-type, probably from further out in the main belt?

Examination of particles within the K-T boundary suggest that it may have been C-type. Moreover, in the mid 1990s F. Kyte found what appears to be a fossil meteorite fragment within the K-T boundary layer of sediments taken from the floor of the North Pacific Ocean. The fragment is just 2.5 mm long and, according to Kyte, appears to be consistent with a Type-3 or Type-2 carbonaceous chondrite. Assuming that this object really is a meteorite fragment and that it is a part of the Chicxulub object, it would therefore follow that this body had a composition similar to that of these meteorite types. In other words, that it was a C-Type asteroid. Furthermore, because the fragment was not very fragile, Kyte saw in this evidence against the cometary nature of the Chicxulub meteorite.

Not everyone is convinced that the fragment is a bona fide meteorite. Moreover, even if it really is a meteorite, the mere fact that it is found in the K-T boundary does not automatically mean that it is a part of the body responsible for that boundary layer. Meteorites fall all the time and some do turn up in ancient sedimentary layers. Kyte's object might simply be an unrelated meteorite that just happened to fall around the same epoch as the Chicxulub body.

Whatever the truth of the matter, the hint that the impacting body was a C-Type asteroid raised an interesting hypothesis.

In 2007, a joint U.S. and Czech team of scientists including W. Bottke, D. Vokrouhlicky and D. Nesvorny, studied evidence of an event—an asteroidal collision—then thought to have taken place within the inner asteroid belt around 160 million years ago. This catastrophic incident, they suggested, may have ejected a number of large asteroid fragments into the inner planetary realm which, in time, became the missiles that impacted both Earth and Moon during the period of interest.

Many of the asteroids of the Main Belt exist in groups or families sharing somewhat similar orbits, a fact discovered as long ago as 1918 thanks to the work of astronomer Kiyotsugu Hirayama, as already noted in the previous chapter. These "Hirayama families", as they are known, arise from the collision of two relatively large asteroids. It was one such family of asteroids that caught the interest of Bottke et al. namely, the so-called Baptistina family, so named for its brightest member, the asteroid 298 Baptistina. According to this team, the Baptistina family is the debris resulting from the impact of an asteroid some 170 km (106 miles) in diameter, by a second body of about 60 km (37 miles) diameter approximately 160 million years ago.

Over time, the Baptistina asteroids slowly spread out through nearby space, with a number of them eventually drifting into a region where the gravitational influence of Jupiter deflected them into orbits crossing the path of the Earth and Moon. The team estimated that as many as 20 % of the asteroid fragments may have ended up in orbits that showed little similarity to those of the core family members. About 2 % of these errant objects are estimated to have struck the Earth.

The time-line initially appeared to agree quite well with the impacts on Earth (Chicxulub and the other smaller craters) and Moon (the Tycho impact) some 60–100 million years later. Moreover, a study of the visual spectrum of Baptistina and another of the brighter members of its family in 2004 indicated that these objects are C-Type asteroids, having a composition similar to that of the Murchison meteorite. This would nicely fit with the proposed Baptistina origin of the Chicxulub meteorite.

However, all did not run as smoothly as might have been hoped. A later study by a team of scientists (including V. Reddy, as well as Bottke and Nesvorny) of the near infrared spectrum of

Baptistina yielded results that were at variance with the earlier ones. Reddy and team found that the near-infrared spectrum of this asteroid showed little similarity with that of a typical C-type. Indeed, even the visual spectrum revealed a weak feature near 0.9 μm that is not present in typical C-type objects, but *is* found in the spectra of S-type asteroids; stony, siliceous bodies similar to ordinary, non-carbonaceous, chondritic meteorites. The new study confirmed this and also yielded an approximate estimate of the object's albedo of between 0.09 and 0.16, making it rather more reflective than a typical dark C-type body.

In spite of these S-type indications, the spectrum did not fall into this class either. It has been listed in the rather broad class of X-type. But the true nature of the asteroid is difficult to determine. Reddy et al. suggest that the nearest meteorite analogues might be CO or CV chondrites (Type III carbonaceous chondrites), the even more highly metamorphosed carbonaceous meteorites known as ureilites or even the non-carbonaceous pallasites. These latter objects may results from shock-induced alteration of rocks brought about in violent collisions between asteroids, whereas the ureilites have, since October 2008, become associated with asteroids of the F-type, a sub-species of the C-types that show a featureless or "flat" spectrum. That month, a tiny asteroid—or, more accurately, a large meteoroid—designated 2008 TC_3 was discovered on collision course with Earth. Fortunately, a spectrum of the object was obtained prior to its atmospheric entry and this identified it as being of the F-type. The object entered the atmosphere over the Sudan and meteoritic fragments subsequently recovered from the Nubian Desert were identified as ureilites albeit with some peculiarities. This dramatic means of identification of ureilites with F-Type asteroids would seem to rule out similarities between the X-Types and these meteorites. If, however, the X-Type reflectance spectrum of Baptistina implies a similar composition to that of a Type III carbonaceous chondrite, which is possible according to Reddy, this could fit with Kyte's possible fossil meteorite—if this really is a meteorite and if it is truly a fragment of the Chicxulub object.

Be that as it may, Reddy and colleagues do not necessarily rule out Baptistina's role in the K-T event, even if its X classification should be found to conflict with a carbonaceous composition.

As the largest member of its family, this asteroid may actually represent the core and innermost regions of the original body and would therefore have been more highly pressure-cooked than the outer parts of the parent asteroid. Even if the bulk of the original asteroid was composed of typical C-Type material, it should come as no surprise that its deeper parts became more highly metamorphosed under these conditions. Any characteristics similar to those of carbonaceous chondrites might have been literally cooked out of the asteroid's core.

It is relevant to note that the Baptistina family overlaps a larger and older Hirayama family named for its brightest member, 8 Flora. The . Flora family members are S-type. There is no debate about this, but because they occupy a well-populated region of the main belt, the family includes a number of pseudo members, several of which are identified as gate crashers by means of their dissimilar classification. The parent asteroid of the Baptistina family can be regarded as having been one of these interlopers; one that gave birth to an entire family of interlopers. Most likely, the smaller body that struck it was one of the . Flora members, and as such an S-type. If this is true, many Baptistina family members may share a mixture of C and S material, making clear classification difficult. Moreover, even today infrared images reveal a band of dust coinciding with the zone of the . Flora asteroids, presumably raised from their surfaces by meteoroid strikes. Surely this stuff is also swept up by asteroids moving through this diffuse stream of dust and gravel. In short, it might be suggested that some of the features detected in the spectra of Baptistina reflect not so much the asteroid's bulk composition but rather the degree that detritus from the. Flora family has accumulated on its surface. Might the 0.9-µm feature be explicable in this way?

These thoughts may no longer be pertinent to the proposed Chicxulub connection. Observations made by the Wide-Field Infrared Survey Explorer (WISE) space observatory and published in 2011 have cast serious doubt upon the proposed association of the Chicxulub meteorite and Baptistina. Infrared observations of a large number of members of the Baptistina asteroid family indicate that the family is a lot younger than formerly estimated. Indeed, the new results measure the age of the family as only about half that of the earlier estimates, that is to say, around 80 million years. This rules out any association with the Copernicus object, but it also casts strong doubt upon association with the objects, including the Chicxulub meteorite, that hit Earth just 20 million years after the Baptistina family itself was formed. For fragments of an object disrupted in the main asteroid belt to reach Earth-crossing orbits, several tens of millions of years of gravitational resonance would typically be required. Put simply, the length of time between the asteroid collision that gave birth to the Baptistina family and the arrival of the Chicxulub meteorite was not long enough for fragments of the Baptistina collision to have come into a position where they could have impacted Earth. While the meteorite may well have been a fragment from the birth of an asteroid family, it now seems a good deal less likely that this family is the one associated with Baptistina.

Periodic Extinctions?

Although the K-T extinction event is the most recent of the major extinctions and the one most widely written about in popular scientific literature, it is certainly far from being the only one which paleontologists have identified. In fact, the transition between the Permian and Triassic eras, around 252 million years ago, was marked by an even greater wave of extinctions in which an estimated 90 % of species and 50 % of genera vanished, leaving the planet bereft of 75 % of its terrestrial vertebrates and 33 % of its marine mammals. Since this catastrophe, some 12 other extinction events have also been noted which, although not as devastating as that which terminated the Permian, still saw the disappearance of many forms of life.

Of the recognized extinction events throughout Earth's biological history, five are considered to have been major. The oldest was the Ordovician-Silurian (O-S) event of 450–440 million years ago in which 27 % of all families, 57 % of genera and between 60 % and 70 % of all species vanished. This episode was really a double catastrophe and is rated as the second largest such event in the planet's history, next only to the great Permian-Triassic extinction.

The second oldest was the Late Devonian (Devonian-Carboniferous) event of 375–360 million years ago. This actually consisted of a series of secondary extinction events spread over an extended period that may have lasted for some 20 million years. About 19 % of all families, 50 % of genera and 70 % of species disappeared during that period.

The next event was the Permian-Triassic already mentioned. This was the most devastating of all the major extinction events and has come to be known, rather ominously, as the Great Dying. Not only did the Triassic era begin with a major extinction, but it also ended with one. The Triassic-Jurassic extinction of 201.3 million years ago saw around 23 % of all families, 48 % of genera and 70–75 % of all species becoming extinct. This event saw the disappearance of most competitors of the dinosaurs on the land, setting the stage for the dominance of these creatures. The age of the dinosaurs therefore began, just as it was to end, intimately related to one of the five major extinction events.

The next event was the one with which this discussion began; the K-T extinction 66 million years ago. Casting their eyes over the geological record of both major and "minor" extinction events, paleontologists J. Sepkoski and D. Raup noted, in 1984, that this record appeared to reveal a periodicity of extinctions having a duration of about 26 million years. Using a larger data base, in 2010 A. Melott and R. Bambach found evidence for a 27-million year periodicity extending back over some 500 million years. Taken at face value, it seemed that something happened every 26 or 27 million years that had a devastating effect upon the planet's ecosystem. But what could that something be?

Sepkoski and Raup suggested a non-terrestrial cause. Some evidence favored this in so far as two such events, the K-T boundary and the Eocene-Oligocene transition 33.7 million years ago each coincide with impact craters of the right age. So it seemed to a number of researchers that impacts by large meteorites might be the cause of these events. But why should these objects arrive periodically and what is the cause of this pattern of extinction events? On the other hand, is the apparent periodicity real or is it simply another of the pseudo-patterns that always crop up in a random scatter of data points?

A number of scientists considered the most straightforward explanation to involve the presence of an object pursuing a very elongated elliptical orbit around the Sun that would bring it, every 26–27 million years or thereabouts, within the outer fringes of the Oort Cloud. A sufficiently massive object in an orbit such as this could cause major gravitational disruption of the comets and asteroids within the affected region of the cloud, sending some careering off into the depths of the Galaxy and others plunging inward toward the Sun. After a time delay, because of the great distances involved, of around two million years, this latter swarm of minor bodies would arrive in the inner planetary system where a certain percentage would very likely collide with the inner planets, including our very own Earth. In the manner of the Chicxulub meteorite, these colliding objects would presumably cause ecological trauma leading to the recorded mass extinctions.

This was the hypothesis of astronomers D. Whitmire and A. Jackson and, independently, of M. Davis, P. Hut and R. Muller. Both teams of astronomers postulated the existence of a small companion star—either a red dwarf or even one of those deuterium-burning sub-stars known as brown dwarfs—orbiting the Sun and planetary system in an elongated elliptical orbit and having its perihelion lying within the Oort Cloud. This hypothetical star was given the name Nemesis or, sometimes more sensationally, the Death Star.

Muller favors the idea that Nemesis is a red dwarf. If that is true, it is a fully fledged star (albeit a feeble one in comparison to the Sun) and should be bright enough to have had its image recorded in photographic star catalogues. Muller suggests an apparent magnitude varying between about 7 and 12. The fainter of these values would still place it within the range of a 6-in. (15-cm) telescope if the user of that telescope knew exactly where to point it. The brighter value would make it an object for small binoculars. Either way, it should have left many images on photographic plates. Nevertheless, there are very large numbers of stars of equal brightness and finding it may not be as easy as we might think. Nearby stars generally give themselves away by revealing a large proper motion, that is to say, they shift their position relatively quickly in comparison to distant background objects and this shift shows up by comparing photographs taken at different dates. However, because Nemesis will be moving either toward or away from the Sun along its elliptical orbit, it will have a significantly smaller proper motion than other nearby stars that are merely passing the Sun on their orbits around the center of the Galaxy. Their motion will be more or less transverse to that of the Sun's and as such they will appear to be moving much faster against the background of more distant field stars. This, plus the relative faintness of Nemesis if it is not at present close to its perihelion, may explain why it had not been picked up on the older proper-motion sky surveys. It would, however, reveal a large parallax, shifting its apparent position as observed at different times of the year, and therefore at different points in Earth's orbit as our planet pursues its own path around the Sun.

Noting that the most recent minor extinction episode occurred 11 million years ago, and taking into consideration the orbits of a number of comets of very long period, Muller derived a possible approximate orbit for the star, indicating that it may now be located in or near the constellation of Hydra. This is a long constellation, so the field in which Nemesis might lie is a large one, but at least most of the search area lies away from the dense Milky Way band!

Further work by Muller in 2002 suggested that Nemesis originally moved in an approximately circular orbit around the Sun, but that this was changed to an elliptical one with an eccentricity as high as 0.7 through an encounter with a passing star some 400 million years ago.

A star in the type of orbit computed for Nemesis would be very open to perturbations by passing stars. Ironically, it is this variation in the orbit of this hypothetical star that caused Melott and Bambach, in their work mentioned earlier, to reject the Nemesis hypothesis as an explanation for the 27 million year periodicity which they claimed to have found in the extinction data. They considered the period too regular to be explained by an object in the Nemesis orbit.

One way in which a red dwarf as nearby as Nemesis might give away its presence is through the tendency of these stars to sporadically give rise to intense flares. Photographic sky patrols in search of transient astronomical phenomena could easily catch such an event. The apparent brightness of the flare would attract the interest of astronomers, who would immediately suspect that either this flare star was abnormally energetic or unusually close and, if a measure of its parallax confirmed the latter possibility, they would become very interested indeed. So far nothing of the sort has happened and all searches for Nemesis have proven negative.

In the face of this, Whitmore and Jackson favor a brown dwarf over a red one. If it is a brown dwarf, Nemesis would be extremely faint visually but it would still radiate in the infrared wavelengths and may have been picked up in the IR scans such as that carried out by the WISE infrared survey. No viable candidate has, however, been found. Indeed, the WISE data pretty convincingly rules out anything larger than Saturn to a distance of around 10,000 AU from the Sun. Larger objects such as supergiant planets and brown dwarfs are ruled out to much greater distances, making it very unlikely that any such bodies are in orbit around the Sun.

In the face of such negative results, the existence of Nemesis is now considered very doubtful. Alternative explanations for the apparent periodicity of extinction events have been put forward, the most interesting one involving an astronomical association involving, not the orbit of an object around the Sun, but the orbit of the Sun around the center of the Galaxy itself. It has been suggested that, as the Sun pursues its galactic orbit, it periodically passes through the denser regions marking out the Galaxy's spiral arms. Although spiral galactic arms make very pretty photographs, they are really quite dangerous places. The very reason why they appear so spectacular in photographs actually betrays one of their dangers. The spiral arms themselves are caused by density waves travelling around the galactic disk. As the wave passes, interstellar material is compressed, triggering a wave of star formation just behind the wave front itself. Among the many stars formed are those with masses several times that of the Sun; massive objects which burn brightly but consume their fuel in only a few millions of years. So brief is their life on the galactic timescale that they do not have time to journey any appreciable distance from the places of their birth. In effect, they die whilst still in the nursery. But because of their great mass, dying for them is not a quiet business. In the words of Dylan Thomas, they refuse to "go gently into that good night", preferring instead to "rage, rage, against the dying of the light". Their death throes take the form of core-collapse supernova which are powerful enough to blast any planet within several tens of light years with lethal doses of radiation. It would not be good for life of Earth if the Sun happened to be passing by one of these cosmic thermonuclear bombs when it went off.

But supernova are not the only danger potentially lurking within the beautiful spiral arms of the Galaxy. Giant molecular clouds have the potential to do exactly what the supporters of Nemesis believed that the death star would do. What with comet showers and supernova, the Sun's passage through the spiral arms of our Galaxy hold little promise for a pleasant interlude.

Actually, considering the dangers posed by these galactic features, it is something of a wonder that we are here at all! The explanation appears to lie in a remarkable and very fortunate relationship between the habitable zone of the Galaxy and what is known as the corotational radius.

The first of these galactic features refers to the region of a galaxy between the dense inner regions, where the situation is at least as hazardous as that within the spiral arms, and the outer fringes where low stellar density means that star formation rates are too low for generations of stars to have synthesized the amounts of heavy elements required for the building of Earth-like planets. The inner city is too rough and the galactic sticks too slow for life to be probable. Somewhere in between—in the outer suburbs of the galactic metropolis—lies the zone where life, other things being equal, has a better chance of taking root and flourishing.

The other feature mentioned above—the corotational radius—refers to the distance from the galactic center where an orbiting star moves at the same velocity as the propagating wave responsible for the spiral arm. The latter does not depend on the distance from the galactic center of gravity and even the motion of stars around the galactic center is not strictly Keplerian in the manner of the orbits of planets within the Solar System. Nevertheless, stars that orbit closer to the galactic center than the corotational radius move faster than the density waves causing the spiral arms and will more or less frequently catch up with these. On the other hand, stars beyond the corotational radius move more slowly than the spiral arms and are periodically overtaken by them. However, if a star happens to orbit close to the corotational radius, and if it lies in between the spiral arms, it can complete circuits of the Galaxy and yet avoid the sorts of dangerous spiral-arm transits that it otherwise would experience. In the Milky Way galaxy, the corotational radius lies within the galactic habitable zone, albeit toward the outer regions of that zone. Moreover, the Sun appears to lie very close to the corotational radius and outside of the main spiral arms, rendering our place in the Galaxy about as safe as it can be made whilst still residing in a region suitable for life! This has not always been so presumably, but it seems to have been this way for a considerable time. Moreover, one of the reasons that it has been this way for a long time no doubt reflects another somewhat unusual feature of our home galaxy; its relatively uneventful history. Like all large galaxies, it has grown by absorbing other neighboring systems, but unlike the majority of large galaxies, the Milky Way appears to have mostly dined on small ones whose absorption into its mass has not resulted in the sort of major disturbances amongst its stars that a merger with a large system would engender. If the Sun's trip around the Galaxy's center is in any way responsible for extinctions, we can be thankful that our remarkable position in the Galaxy is such as would keep these to a minimum, at least throughout recent cosmological epochs.

Even though the Earth has apparently been kept from too high a frequency of transits with galactic spiral arms, crossings have occurred and at least some of the extinction episodes may have been caused by the violent phenomena occurring within these regions. But apparently that is far from being the whole story. More recent work has called into question the reality of the periodicity itself, at least, as being anything more than random statistical fluctuations. If this is correct, there is no longer any reason to search for a single underlying cause of the extinction episodes. Some could have one cause and others an entirely different cause!

In apparent support of this, it seems that, whereas the K-T event coincided with a major impact, the Triassic-Jurassic extinction 201 million years ago had no astronomical associations. On the contrary this event appears to have been triggered by an outbreak of mega-volcanism associated with the Central Atlantic Magmatic Province during the separation of the North American continent from northern Africa as the ancient Pangean supercontinent split apart in the giant rift that formed the Atlantic Ocean. During the course of 600,000 years, over three billion cubic kilometers of basaltic lava were produced in four pulses while carbon dioxide and other gases were belched into the atmosphere and dissolved in the waters, turning the oceans dangerously acidic.

While admitting that some extinctions may indeed have been triggered by a rain of comets from the Oort Cloud disturbed from their habitat by a passing ,star or a passage through the galactic arms, the weight of evidence (such as it is) is not in favor of this cause for the K-T event. Even though the impacting object may not have originated in the Baptistina family, it remains a good bet that it was an asteroidal body originating somewhere with the main belt. If that is correct, it was quite unconnected with anything that may have happened in the Oort Cloud.

A Star Passed By

Although Nemesis may not exist, a star having very similar characteristics has recently been discovered. Moreover, backtracking the motion of this star through time revealed that it did indeed make a very close approach to the Solar System just 70,000 years ago. The close approach was discovered by astronomer Ralf-Dieter Scholz and announced in November 2013. The star is currently some 20 light years distant, plus or minus 3 light years, but passed as close as 0.8 light years from the Sun (and therefore inside the outer regions of the Oort Cloud) at the time of its minimum approach. Recalling the difference of opinion as to whether the hypothetical Nemesis might be a red or a brown dwarf, this real life "Nemesis" solves the debate by being both. It is a double star; a red dwarf with a brown dwarf companion! At its closest approach, it is estimated to have shone with an apparent magnitude of about 11 but, typical of red dwarfs, it is prone to sudden brief flares and may have spiked to naked-eye visibility on occasions during its visit (Fig. 3.7).

Did Scholz's Star Cause a Comet Shower?

It probably did, however the disrupted comets are so far from the Sun that they are still on their way. They will not reach the planetary system for nearly two million years, at which time the number of incoming dynamically new comets will probably peak dramatically, although probably not enough to endanger our distant descendents. It is estimated that, on average, a star similar to Scholz's passes by at comparable distances about once every nine million years. Placing these close approaches in perspective, if the distance between Earth and Sun is represented by one inch and

FIG. 3.7 Artist's impression of Scholtz's Star and its brown dwarf companion (*Courtesy*: Michael Osadciw, University of Rochester)

the Sun by a pin's head, Scholz's star would have passed some four fifths of a mile away on that same scale—a very close approach.

But Are Comet Showers Necessarily Dangerous?

From a reading of popular accounts of extinction events, comet showers and explosive impacts, it is easy to gain the impression that we are living, not so much at the wrong end of a shooting gallery as on the target of a bombing range. Certainly, there is an ever-present danger from the skies, but it must not be overestimated. We should not succumb to meteorite phobia.

The Earth shows quite a number of scars acquired over millions of years and, popular accounts have made much of the one that coincides with the mass extinction at the K-T boundary. Yet, there is little mention of the non-correspondence between most of the large meteorite craters and signs of global ecological trauma. True, the majority of the craters blasted out since our planet acquired a complex biosphere are not anywhere close to the size of Chicxulub, although the Kara Crater in Russia is not greatly smaller (120 km compared with Chicxulub's 180) and is estimated to have been formed by an impact just over four million years earlier. Of the smaller craters, central Australia's Gosse's Bluff, at 22 km (14 miles) diameter and formed about 142 million years ago dates to a time close to the Jurassic-Cretaceous boundary, however that particular boundary is not marked by any extinction event. It seems that impacts, even quite large ones, do not necessarily result in major trauma to Earth's biosphere. Perhaps there is a critical size above which major trauma becomes likely and maybe that size lies somewhere between that of the Kara and Chicxulub events. Or maybe the global results depend upon where on the planet the asteroid strikes. Or, alternatively, it is possible that a major extinction episode depends upon more than just one event. Perhaps the event singled out as the trigger (whether an asteroid collision, a mega-volcanic episode or anything else) was just the final straw that, as the saying goes, broke the camel's back. As mentioned earlier, there is some thought that the dinosaurs were already on their way out before the Chicxulub meteorite struck. Maybe the entire Cretaceous ecosystem had passed its prime and that it was only a matter of time before something happened to kick in the already rotting door. If it had not been an asteroid collision 66 million years ago, it might have been something else happening in more recent times. Or the system may simply have broken down over millions of years in a slow transition into the Tertiary era (Fig. 3.8).

The other popular notion which may need modifying is that of the comet shower. Popularly, these episodes are thought of as times when the night sky fairly blazes with comets. Artists' impression typically depict a sunset or dawn scene with two or three great comets sporting their tails in the twilight and a further two or three fuzzy blobs higher in the darkening sky. Such scenes probably did occur when the Solar System was very young, but how realistic are they for the typical comet shower arising from the average close stellar encounter? How well will they represent, for instance, the coming comet influx resulting from the close passage of Scholz's Star? Most likely, not very well. In fact, there is some indication that we might be living through a comet shower right now.

The Oort Cloud, the vast region of the outermost reaches of the Solar System populated by immense numbers of small objects,

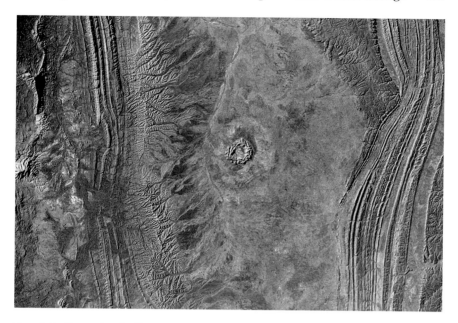

Fɪɢ. 3.8 Gosse's Bluff—an ancient impact crater imaged from the International Space Station (*Courtesy*: ISS, NASA)

principally of a cometary nature. The objects in this cloud are too remote to be visible from Earth, but there are very good reasons for believing that the cloud is real and its presence plays a pivotal role in the concept of comet showers. Back in the middle years of the last century, Dutch astronomer Jan Oort found something curious about a sample of comets of very long period that he was examining. There was a significant grouping of the semi-major axes of their orbits at distances so far removed from the Sun as to be barely susceptible to its gravitational pull. It was this odd grouping that led him to hypothesize the existence of the comet reservoir now named the "Oort Cloud" in his honor. Oort figured that the overwhelming majority of these objects must be making their very first close approach to the Sun (close in the sense of passing it at distances comparable with those of the planets). So tenuous is the Sun's hold on these objects that even a slight perturbation by one of the planets would be enough to either boost the comet's orbital eccentricity to the escape velocity of the Solar System or else retard it enough for the comet to begin falling back toward the Sun well before it had again receded to the distance of its initial aphelion.

Because the orbits of these objects began so close to the parabolic limit dividing hyperbolic and elliptical orbits, only a slight boost or tug would be required to achieve either ejection into interstellar space or retardation into an ellipse of very long period. The majority would end up in ellipses of long period. Whereas about half of these objects gain slight hyperbolic eccentricity while relatively close to the Sun, at great distances many of these hyperbolic orbits revert back into ellipses under the gravitational influence of the Solar System as a whole. While a few escape into the depths of the Galaxy, other hyperbolic comets end up in surprisingly short period orbits; tens of thousands rather than hundreds of thousands of years, although there are also those that settle into ellipses with periods of a million years or more.

The upshot of this is that there should be large numbers of second-time-around comets coming into view along orbits which, while still very eccentric and having periods ranging from several tens of thousands to hundreds of thousands of years or even a million years and over, nevertheless fall short of Oort Cloud distances. It would be expected that their numbers will be somewhat less than the truly dynamically new ones given that some of these will end up in truly hyperbolic orbits and very small and/or fragile ones are known to disintegrate, but the discrepancy should not be too great. The surprising thing is that the discrepancy is great—so great in fact that Oort concluded that most comets must fade dramatically after their first approach to the Sun and that only around 10 % remain bright enough to be discovered at their subsequent apparition.

If this is true, it implies that, whereas some 90 % of comets are fragile bodies that don't survive their first experience of activity near the Sun, the remaining 10 % are, for the most part, durable objects that return again and again with little diminution in their light even after many apparitions. Why this strange gulf with little gradation between the fragile majority and the durable minority?

The direct observational evidence for such a large attrition rate is not at all obvious. There is indeed a strong tendency for small and intrinsically faint dynamically new comets, especially those venturing within the Earth's orbit, to break up and fade away, but they certainly comprise a lot less than 90 % of the population. There may be further loss in so far as dynamically new comets activate at unusually large distances from the Sun thanks to a thin coating of very volatile substances on the surfaces of their nuclei. This material causes the comet to display activity when it is still too far from the Sun for water ice to sublimate and most known comets that have their perihelia at solar distances of several Astronomical Units do in fact turn out to be dynamically new. These objects probably lose their volatile coating on their first approach to perihelion and, assuming that their perihelion distance remains constant between returns, they are likely not to be active at their subsequent return. Their very volatile coating will be gone and, as they do not come close enough to the Sun for water ice to drive activity, they presumably either remain inert or only display weak activity. If this line of thinking is correct, some of the attrition rate is explained, although still not enough to account for the extent of the drop in numbers. Indeed, this effect would have only a minimal effect on Oort's original sample as few comets of really large perihelion distance had been discovered at the time of his research.

It might also be worth mentioning that E. Opik, thought that a dynamically new comet would have its orbit altered by passing stars before it had a chance to return on a second visit to the Sun and, in its new orbit, likely remain too far away to be seen again. Although he was not addressing the problem of the gap in numbers between first-time and second-time comets (of which he may not even have been aware at the time of his writing) his opinion that dynamically new comets mostly pass this way but once would solve the problem if confirmed. Unfortunately though, the effect of passing stars will, at most, only affect a small percentage of objects. The majority—the ones that end up in orbits extending to distances significantly smaller than that of the Oort Cloud—will be immune to the influence of passing stars and will, in any case, have orbital periods short enough to have covered their orbits several times between the stellar encounter that deflected them in the first place and any subsequent ones.

There seems to be only two likely explanations for the gap in numbers. The first, in agreement with Oort's conjecture, involves the majority of dynamically new comets slowly crumbing away as they retreat from the Sun following their maiden perihelion passage. This is not what we observe. True, some comets do fade out under observation. True again, this tendency is greater for dynamically new ones than it is for those arriving on ellipses of long period. But these objects are nevertheless still in the minority. If fading and disintegration is the answer to the relative paucity of second-time comets, many of the dynamically new objects that appear to remain healthy must succumb once activity ceases and they fade from view. This behavior appears strange to me, although it is certainly a possibility.

The second alternative is to give up the implied assumption that the observed influx of dynamically new comets is the steady-state one. In other words, the problem is solved if we currently are experiencing a comet shower. If that is true, the second-time comets that we are discovering today better reflect the true steady-state rate of dynamically new ones during non-shower epochs, once corrected for the minority that disintegrate, deactivate or are ejected from the Solar System.

If we are experiencing a comet shower, two consequences immediately present themselves. First, the estimated population of comets residing in the Oort Cloud that is most often quoted (10^{11}) will be too high, probably by a factor of 10 or so and, secondly, most comet showers are not the devastating affairs that they are typically presented as being. A contemporary comet shower also implies that there was a stellar encounter around two million years ago. So although the average passage as close as that of Scholz's star is around nine million years, it may be that there have been two such encounters in the (cosmically speaking) relatively recent past.

Of course, there is always the possibility of a comet hitting Earth. This has happened in the past and will surely happen again in the future, but the greater danger is almost certainly from asteroids in short period orbits having low inclination to the plane of the ecliptic. While we cannot rule out cometary impacts as the cause of (or as a contributing factor to) some of the extinction events, there is no compelling evidence that this is the case. Furthermore, the relatively major meteoritic events of recent history (for example, Tunguska 1908, Sikhote-Alin 1947 and Chelyabinsk 2013) all had asteroidal bodies as the culprits. To our knowledge, only one tiny comet is thought to have hit the Earth in recent years; the fireball that lit up the skies over Northern California early in the morning of 17 January 2013. This was a fragile body moving in an orbit that apparently came in from the Oort Cloud. However, the object was very small and dissolved high in the atmosphere without causing any damage. We do not know if it was active (if it possessed a small coma) but even if it had missed Earth, it was almost certainly too small to have survived its passage around the Sun.

4 Comet Controversies

Are Comets Ejected by Planets, Their Moons … and the Sun?

Even if comets may not be guilty of mass extinctions, they have certainly been the subject of some hypotheses of varying degrees of weirdness. Among twentieth century researchers into comets and associated phenomena, the name S.K. Vsekhsvyatskii looms large. His analysis of the changing brightness of comets as they approach and recede from the Sun enabled him to estimate an average rate of response to the changing degree of solar radiation which is still widely used in making brightness forecasts for these objects, even though more recent alternative values are increasingly taking over from what had long been considered as Vsekhsvyatskii's standard. Nevertheless, his research into the way comets respond to their distance from the Sun resulted in one of the most comprehensive descriptive catalogues of historical comets available until recent years, published in his book *Physical Characteristics of Comets* and later supplements, which has proved invaluable both to researchers of comets and to historians of astronomy.

His ideas as to the origin of these objects became increasingly at odds with the wider development of thought on this subject during the latter half of last century. Vsekhsvyatskii championed the planetary ejection hypothesis of cometary origins. During his earlier years, he held the view that comets were ejected from the major planets. Jupiter was thought to be the principal comet factory of the Solar System, as is apparently evidenced by the quite large family of these objects having the aphelia of their orbits close to the orbit of the giant planet. Saturn seemed to have a comet family as well, albeit far smaller than that of Jupiter, and Uranus and

© Springer International Publishing Switzerland 2016
D. Seargent, *Weird Astronomical Theories of the Solar System and Beyond*, Astronomers' Universe, DOI 10.1007/978-3-319-25295-7_4

Neptune were each thought to sport a very few associated comets. Vsekhsvyatskii was not the first to propose the planetary ejection theory. It had already been suggested by J.L. Lagrange as early as 1814, again by R. Proctor in 1870 or thereabouts and yet again by F. Tisserand some 20 years later. Nevertheless, Vsekhsvyatskii's research into cometary brightness raised issues which breathed fresh life into the hypothesis, as we shall soon discover.

Majority opinion read something entirely different into the existence of comet families associated with the giant planets. These planets had cometary families, not because they somehow spat them out of their bulk into surrounding space, but because their powerful gravity captured objects that were entering the inner regions of the Solar System from far more distant fields. The hypothesis independently arrived at by Opik and Oort was becoming increasingly favored. Nevertheless, despite the general success of this model, it faced a serious difficulty in explaining the nature of the system of short-period comets. Because the Oort Cloud should be approximately spherical, new arrivals from this region necessarily arrive on orbits displaying no preference for the orbits of the planets. They come in from every possible direction without any regard for the ecliptic plane and their orbits are either slightly hyperbolic or ellipses of such length as to be almost indistinguishable from parabolas and having aphelia far, far, beyond the furthest planet. Yet, the accepted theory held that it was from these objects that Jupiter and the other giant planets conscripted the members of their families.

Precisely because most of the Oort-Cloud comets pay no heed to the orbital plane of the planets, they are not readily amenable to capture into the sort of orbits that we observe their short-period brethren to follow. Moreover, capture of a comet coming in along an orbit of extreme eccentricity is itself no easy matter. And to make matters even more difficult, the number of known comets belonging to Jupiter's family just kept increasing as large-field photographic surveys directed toward stellar proper motion studies, star charts and minor planet data kept turning up new comets of both short and long period as a by-product of these projects. The number of known comets belonging to Jupiter's family grew too large to be explained by the inefficient method of capture from the field of long-period objects—and with the number of new discoveries increasing all the time, the situation was rapidly getting worse.

The problem was further aggravated by the relatively short lives of comets in the Jupiter family. Short lives actually refers to two quite separate things. The most straightforward meaning is that because a comet loses some of its mass, in the form of gas and dust, each time it approaches the Sun and becomes active, there must come a time where it simply has nothing left to give. It either disintegrates completely or else turns into a dormant asteroidal type of body devoid of the volatile substances that previously caused it to sprout a visible coma and tail. Most likely, some comets break up completely while others leave an inert cinder, but either way, they must eventually fade from view. The only way of avoiding this fate is for the comet to have its orbit so altered by a close passage to one of the planets (almost inevitably, Jupiter) that it is either expelled from the Solar System altogether or has its perihelion distance so enlarged that even at its closest approach to the Sun, it no longer receives sufficient warmth to become active. In this way, a comet might have its physical life preserved, albeit at the cost of what might be called its active dynamic life. It continues to exist and still retains a large enough store of volatiles to become active if it should again be exposed to sufficiently high temperatures, but it's orbit has been so altered that it remains dormant. Whether a comet reaches the end of its physical or simply of its dynamical life, the result is the same, observationally speaking. The comet is lost. Because of their short periods and consequent regular bouts of activity, plus the frequent gravitational perturbations of giant Jupiter, the members of Jupiter's comet family cannot have more than very brief lifespans on the cosmological scale of things. If they manage to avoid complete loss of volatiles, it is only at the expense of being shot out of the Jovian family altogether. There seems to be no escaping the conclusion that the comets we see in the giant planet's family today will not be around in the not-too-remote future and were not present in the not-too-distant past. Assuming that Jupiter's present contingent of comets is not some rare fluke which our own age is specially blessed to witness, multiple generations of comets must have come and gone over the age of the Solar System. This once again highlights the efficiency of the process endowing Jupiter with its hordes of comets.

The actual lifetime of these comets is a matter of debate. Everyone agreed that it must be short by the standards of the age of the Solar System, but just how short? Maybe a few thousand years? That certainly would be brief considering the age of planets. Assuming that the capture process did manage to work, that meant that once a formerly dormant comet was captures by Jupiter, its future active life would be reduced to a timescale more normally applied to human societies than to astronomical objects.

Yet even this still length of time seemed too long for some astronomers and it was amongst these that Vsekhsvyatskii was to be numbered. Thanks to his laborious work in compiling his catalogue and deriving the absolute magnitudes (i.e. the hypothetical brightness that a comet would have if located at one Astronomical Unit from both the Sun and the observer) of every comet that had been sufficiently well observed, he was able to compare the brightness of those periodic comets visible in the nineteenth century with the same objects as observed during the following century. Remarkably, he found that the absolute brightness of most of these had faded by at least a magnitude—significantly more in some cases—during the previous hundred years or thereabouts. From this, he concluded that the active lifetimes of most short-period comets are very short indeed; measured in terms of a few centuries at most and reduced to just decades in the more extreme instances. Combining this evidence for very short lifetimes with the relatively large contemporary number of these objects and the inefficiency of the capture process as conceived at the time of his research, Vsekhsvyatskii saw no alternative to the hypothesis that comets were ejected from the giant planets themselves. Comets having extremely elongated orbits were also ejected from the planets, but in their instance the process of ejection was so violent that they acquired sufficient velocity to escape into the wider realms of space. Vsekhsvyatskii supposed that Jupiter had settled down somewhat since its tempestuous youth and that it is now unlikely to experience eruptions powerful enough to eject objects having velocities sufficiently in excess of its escape velocity to end up in very elongated ellipses. The other outer planets with their smaller escape velocities were therefore more likely responsible for most of the long-period comets that we see today. Thus, rather than the general field of long-period comets supplying the planets with those of short period, as envisaged by the capture hypothesis, the situation was effectively reversed. Long-period comets were also born from the planets, but managed to escape thanks to the extra violence accompanying their nativities.

Nevertheless, the theory came with some serious problems of its own. The principal objection was the sheer violence required for any process to eject an object the size of a comet from the powerful gravitational field of Jupiter. Saturn and the other giant planets did not pose quite as big a problem, although the concept of this sort of super volcanism was not easy to comprehend, especially as nothing of the sort had ever been observed taking place on any of these worlds. It may be supposed that if comets are being ejected from Jupiter, in particular, at a rate required to replace the rapidly-dissolving members of its cometary cohort, something would have been witnessed by now. Moreover, considering the deep and dense atmosphere of this world, a problem surely arises in understanding how a comet could be ejected upward through this mantel of gas without being destroyed in the process. It would be like a comet/Jupiter collision played backwards, with equally destructive consequences for the smaller of the two bodies.

Influenced by this line of thinking, especially the question of excessive velocities being required for an ejected comet to escape Jupiter, Vsekhsvyatskii later modified his thesis by placing the region of ejection, not on Jupiter itself, but on that planet's major moons. As the escape velocity of these is hardly even comparable with that of their primary far less energy is required to shoot a comet out into space. Moreover, because these moons possess little by way of atmosphere, such an ejected object would not encounter the destructive resistance that it would experience if fired off from deep within Jupiter itself.

This shift in location of the hypothesized launching pad of comets led Vsekhsvyatskii to make certain predictions concerning Jupiter's moons and environs. He predicted that, because the ejection of a comet would likely also entail a certain amount of accompanying debris (indeed, he argued that at least some meteor streams can be thought of in this way rather than wholly as trails of particles shed by the comet itself) a certain amount of particulate debris has probably gone into orbit around Jupiter. Like Saturn, this planet should have a ring system. When he made this prediction, it should be recalled, Saturn's ring system was thought to be unique, at least in our own Solar System. Moreover, because comets are ejected, according to his thesis, from Jupiter's moons by some form of volcanic activity and because this process must be a continuing one (required by the short lifetime of Jupiter's comets) at least some of these moons must currently display active volcanism. Once again, this was a radical prediction made in an era when the only confirmed volcanism within the Solar System was on our own planet.

But what did continuing research, especially from space probes to the outer planets, reveal in later years? A ring around Jupiter and a volcanic situation on Jupiter's moon Io that puts anything Earth can bring forth to shame. Vsekhsvyatskii's hypothesis made two successful predictions, both of which were quite out of left field at the time he made them. Does that success mean that his theory has now been proven correct?

The short answer to this question is "No". But the scenario raises some interesting issues of a philosophical nature concerning when and why a certain theory can be regarded as having been proven correct, as well as the nature of proof itself. Incorrect theories can make valid predictions, which are traditionally evidence that a theory is correct. So why, in the face of Vsekhsvyatskii's valid predictions, can this particular theory of his not be accepted?

Essentially, his theory fails to convince because it encounters too many difficulties and because most of the data upon which it relied has been proven false. Moreover, the apparently difficult facts which initially seemed inexplicable on the basis of any alternative theory have since proven to be less difficult than they at first appeared. When the evidence that initially seems to require the validity of a particular theory is weakened by further research, such that less controversial theories begin to look equally viable as explanations, there is a good chance that the more straightforward theories will prove more successful in the end. This is just what happened in the case of Vsekhsvyatskii's model.

Although Vsekhsvyatskii's switch from Jupiter to the planet's moons as the source of comets certainly made his theory more credible, serious difficulties nevertheless remained. True, the discovery of volcanoes on Io seemed to verify one of the theory's predictions, but did their discovery really help explain the genesis of comets? Comets have picturesquely been called dirty snowballs and the majority of them are quite fragile bodies. The astronomer who said that if one were to sneeze on a comet it would break up might have been exaggerating (although only slightly in some instances!) but the point was nevertheless well made. So just how could a fragile icy body be ejected from one of Io's volcanoes without being wrecked in the process? In fact, how could something composed largely of water ice be ejected from Io at all, in view of the dry nature of this moon? And the wettest of Jupiter's moons, Europa, does not possess the level of volcanic activity required for the job.

Furthermore, nothing even remotely resembling a comet has been observed to usher forth from any of Jupiter's moons, not even during the most violent of Io's volcanic outbursts. Neither has a new comet suddenly appeared close to this or any other moon, so far as we are aware. As far as the predicted ring around Jupiter is concerned, this can quite readily be explained without the need for volcanism on the scale of violence required by Vsekhsvyatskii's thesis.

The same can now be said about the relatively large number of comets belonging to Jupiter's family. Caught between arguments for the short lifetimes of these objects and the difficulty of their capture from the wider field of long-period comets, the theory of Vsekhsvyatskii appealed as a neat way through the dilemma. However, as continuing research brought new facts to light, both of these apparent obstacles grew less menacing. Capture of comets arriving from the Oort Cloud may have remained as difficult as ever, but a new and much more amenable reservoir of suitable objects was found in the form of the Kuiper Belt; a disc-shaped population of comets and other minor bodies beyond the orbit of Neptune. Because the Belt is a lot closer than the Cloud and because its denizens pay more heed to the ecliptic plane, objects originating in that region can readily migrate inwards under the action of gravitational perturbations by the major planets. For a long time, comets coming in from the Kuiper Belt reside in Centaur orbits, as, for example, the giant cometary bodies Chiron and Echeclus. But they cannot remain in these orbits forever. Thanks to the gravitational perturbations by the giant planets, they will eventually either be expelled from the planetary region or else brought in closer to the

Sun where they can be captured by Jupiter into the typical orbits followed by Jupiter-family comets. This capture process presents no serious problems. If the large Centaurs that we observe are just the tip of an iceberg of great numbers of smaller objects—objects having the dimensions of typical short-period comets—there are no problems in accounting for the number of these objects and an important hurdle to the capture hypothesis has successfully been removed.

It is also interesting to speculate that the present population of Jupiter-family comets may also contain a high proportion of fragments from a large Centaur that broke up (possibly because of a major collision with another object?), forming a belt of hundreds of thousands of diminutive Centaurs, some of which ended up as Jupiter-family comets. Although the population of the Jupiter family does not require this scenario, it does remain an interesting possibility. At different times in Solar System history, large bodies of the type exemplified by Chiron, Echeclus and Saturn's moon Phoebe have surely migrated inward and become Jupiter-family comets. We might speculate that one of these would now be in such an orbit had it not met with disaster whilst still a Centaur and that the Jovian comet retinue that we see today are some of its fragments. Maybe not all of Jupiter's family, but perhaps a large percentage of these objects may have derived from a common parent. The improbability of all the present retinue of Jupiter's comets descending from a common ancestor is implied by the difference in the deuterium to hydrogen ratio between the Jupiter-family comets 103P/Hartley and 67P/Churyumov-Gerasimenko. The former was found to have a D/H ratio similar to that of the water found in carbonaceous meteorites and in Earth's ancient water whereas the D/H ratio of the latter is more in keeping with comets of longer period. Which (if either) is more representative of Jupiter's family remains to be determined.

As a wild speculation we might wonder if this implies that the parent Centaur (if there indeed was one) broke up following a collision with a comet from the Oort Cloud. On the other hand, the D/H ratio of the plumes emitted by Saturn's moon Enceladus is also closer to that of a long-period comet, so if there is any truth at all in our suggestion that today's short-period comets are fragments from a collisionally disrupted Centaur, potentially plus fragments from the colliding body itself, the colliding body may have had a D/H ratio of a long-period comet and yet still moved in a Centaur-like orbit. A collision between two Centaurs, though improbable enough, is still more likely than one involving a Centaur and a comet from the Oort Cloud.

Do Jupiter's Comets Quickly Fade Away?

Be that as it may, further research has likewise largely dispensed with the other big hurdle for the capture theory, namely, the short lifetimes of Jupiter's comets. As each of these hurdles erected in the path of the capture theory equally acted as supports for Vsekhsvyatskii's thesis, the latter naturally lost its appeal as the former gained strength.

Concerning the average lifetimes of Jupiter's comets, no-one is denying that certain members of the Jovian family have faded and there is no disputing that several have disappeared altogether, presumably breaking up and disintegrating just as Vsekhsvyatskii envisioned. Yet others have had their orbits so altered that they no longer approach as closely to the Sun and Earth and have become invisible and, maybe, inactive for want of solar warmth. But the disintegrating comets were either old objects already nearing the end of their lives by the time of their discovery or else they were atypically small and/or fragile bodies that simply could not last the distance. For the most part, the apparent fading that Vsekhsvyatskii believed he had found and on which he based his thesis of the very rapid decay of short-period comets has not been supported by more recent studies.

There are several reasons why Vsekhsvyatskii was mistaken about this matter. For a start, a number of short-period comets had been unusually bright at their discovery apparition. Indeed, it is this that brought about their discovery. Sometimes this extra luster has been due to one of the outbursts that some comets experience from time to time. Several comets have also been discovered shortly after their orbits had been drastically altered by a close passage of Jupiter, such that they came to approach the Sun far closer than they had during their previous (unobserved) returns. Quite often, a comet will appear extra lustrous on its initial passage—

or even during its first two or three passages—through a smaller perihelion distance, presumably because previously undisturbed deposits of ice on its surface are boiled away at their first experience of significant solar heating. Although these cases might be cited as instances of cometary fading, they are not the sort of fading that is synonymous with decay. It would be better to think of these comets at their initial return(s) as displaying an anomalously high brightness which returned to normal during later appearances.

There is, however, a more general reason why periodic comets appeared to fade between the mid nineteenth and mid twentieth centuries; and it has nothing to do with decay of the comets themselves. The reason lies in the way in which they were observed in earlier years. Back in the mid 1800s, astronomers observed comets visually using telescopes of what were, by today's standards, of rather small aperture. Because of the modest aperture of these instruments, the magnifications employed were also low or moderate. Now, paradoxical though it might seem, when the typical comet is observed under high magnification, it appears smaller than when seen through a wide-field telescope at low power. The reason for this odd result lies with the diffuse nature of these objects. Typically, a comet appears in a telescope eyepiece as a misty ball with a brighter and more intense core, surrounded by a thinner luminous haze which becomes decreasingly intense with distance from the center. When observed with low power, a comet appears as a more or less concentrated fuzz ball within a comparatively wide field. Somebody making a brightness estimate will be seeing all of the comet's light concentrated into a relatively small image. But switch to a high magnification and the outer regions of the comet become so diffused through over magnification that they fall below the threshold of visibility and are no longer included in the brightness estimate. In effect, someone using a large high-magnification telescope only estimates the inner region of the comet and, in consequence, the estimate will be fainter than that of another person estimating the same object with the help of a pair of binoculars—other things, as always, being equal. But the situation became even worse when photography became the principal tool for observing. In extreme instances, brightness estimates of the same comet can differ by over 250 times depending upon whether the observer is using photography or a small low-magnification visual telescope. As photographic magnitude estimates increasingly came to dominate the observational records, the impression was given that the comets themselves were fading. This became a catch-22 situation; observers armed only with small telescopes looked at the reported brightness estimates coming from the big observatories and decided that they had no hope of seeing the comets with their meager instruments. So they didn't try to observe, thereby leaving the field dominated by the large photographic instruments.

Doubts about whether the comets were truly fading started to arise in the late 1970s. Based upon their supposed intrinsic fading, a number of short-period comets had been predicted to become defunct around the middle of last century. Yet, they continued to turn up and, worse still for such predictions, several that had been lost for years were recovered following the application of new computer technology to the determination of their orbits.

One such object was Pons-Winnecke. This comet had an interesting history. When discovered by J. Pons in 1819, it came to perihelion just beyond the orbit of Venus. The same was true when it was rediscovered by F. Winnecke in 1858. However, the comet was locked in a 2:1 resonance with Jupiter, which meant that every 12 years it would experience a relatively close encounter with the giant planet and the cumulative effect of these encounters resulted in a steady increase in the comet's perihelion distance. By the end of the second decade of last century, its perihelion lay close to the orbit of Earth, allowing for close approaches to our planet in 1921, 1927 and 1939. By the mid-1940s however, the comet's perihelion lay outside of Earth's orbit and by the early twenty-first century, it had increased to nearly 1.26 AU, before slowly starting to shrink again (it was down to 1.24 in 2015). Pons-Winnecke was a bright telescopic object during the nineteenth and early twentieth centuries. Indeed, it was clearly visible with the naked eye during the very close approach to Earth in 1927. However, with the increasing perihelion distance, it stands to reason that a certain fading should have taken place as the comet withdrew from the hotter regions of the Solar System and, presumably, became less active. Nevertheless, according to observational accounts, it faded more than could reasonably be expected from this process alone. It had apparently gone faint from the end of the 1930s and was missed altogether at the predicted return of 1957. Assuming that this greater-than-expected apparent fading of the comet was indicative of its general decay, well known comet expert Fred Whipple, the man who solved the mystery of the nature of comet nuclei, predicted a complete fade out by 1960. The missed return of 1957 appeared to confirm this forecast (Fig. 4.1).

It was with some surprise therefore that Dr. Elizabeth Roemer of the US Naval Observatory recovered the comet at its 1964 apparition. Clearly, the comet was still alive but Roemer's estimates of its brightness (made, of course, photographically with a large telescope) placed it at a faint magnitude 17 at maximum; a far cry from its former performances, even allowing for its greater minimum distance from the Sun.

The next two returns of this comet were similarly faint, but as the return of 1983 approached, Dutch comet observer Reinder Bouma decided to try an experiment. By the early 1980s, several amateur astronomers had begun to deliberately seek out returning comets of short period, using similar instruments to those used by astronomers of the previous century. Increasingly, these objects were observed and found to be far brighter than the (photographically based) official brightness predictions indicated. A co-operative effort by Dutch and Australian amateur astronomers, put into place by Bouma and the present writer, meant that suitable periodic comets could be monitored from both hemispheres. Noting that the 1983 return of Pons-Winnecke would be favorable for southern hemisphere observers, Bouma alerted the Australian Comet Section to the possibility that it might be a lot brighter than predicted. He reasoned that during recent returns, the comet had been only observed photographically by means of large telescopes and that the published magnitude estimates were, as a consequence, likely to be much too conservative. The catch-22 situation mentioned above then came into play; visual observers concluded that it had intrinsically faded and would no longer be visible in their telescopes, and therefore failed to even try seeing it. But, according to Bouma, if the lion's share of the fading was due to the suggested instrumental artifacts, and the rest was simply a consequence of the comet's changing orbit, there probably had been no appreciable decay and the comet might still come

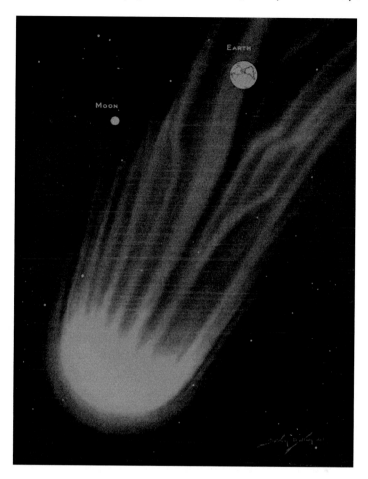

FIG. 4.1 An artist's rather fanciful impression of Comet Pons-Winnecke during its Earthgrazing encounter of 1927 (*Courtesy*: Wikimedia)

within the range of small telescopes. I confess that I was skeptical, but circumstances proved Bouma to be completely correct. Western Australian amateur astronomers Maurice Clark and Andrew Pearce found the comet glowing softly between magnitudes 12 and 13 and, a little later, American observer Alan Hale (later to become co-discoverer of the huge Comet Hale-Bopp), after having been alerted by the writer, also observed it during his visit to Australia on his way home from viewing that year's total solar eclipse in Indonesia.

It is also worth mentioning that the comet was seen in 2002 under very similar conditions to those 19 years earlier and once again reached twelfth magnitude in small telescopes. Clearly, this object is not in the process of rapid decay. In view of results of this nature, the urgency of replacing Jupiter's store of comets on a very short timescale through the ejection of these objects from either the planet itself or its moons is taken away.

Although we have been concentrating on the Jupiter family, Vsekhsvyatskii also thought that the smaller comet families of the other outer planets were likewise the product of eruptions on the moons of these worlds. These families are very much smaller than that of Jupiter. In fact, the supposed families of these planets now seem to be more apparent than physically real. Comets having periods longer than those of Jupiter's family pay less heed to the ecliptic plane and what looks like a close approach to a planet in orbital diagrams drawn on the two-dimensional surface of a sheet of paper can take on an entirely different appearance if presented as a three-dimensional model. A striking example of this is the apparent relationship between Halley's Comet and Neptune. This comet has been claimed as a member of that planet's family and, indeed, it appears this way on a flat representation of the orbits of these two objects. However, in actual fact the orbit of Halley is strongly inclined to that of Neptune and the other planets and the comet actually comes nowhere near the planet. At its closest, Halley is still some eight AU distant from Neptune and experiences only slight perturbations from the planet's gravity. It comes a lot closer to Jupiter however, which has a considerably stronger influence on its orbital motion. In short, the likelihood that Halley's Comet could have been volcanically ejected from Neptune or one of its moons is just about zero.

Solar Comets?

Whilst speaking of comets being ejected from other astronomical bodies, mention should also be made of a strange speculation made by one of the recognized experts of cometary astronomy, A.C.D. Crommelin. Crommelin suggested that these objects may have been spat out of the Sun and that in some way this ejection process was associated with solar prominences. At the time this suggestion was made, over a century ago, little was known about the nature of solar prominences and ideas about the nature of comets were quite unlike those of today. Yet, granted this, the suggestion still strikes one as a little weird. Professor R.A. Lyttleton, writing around the middle years of last century, made the rather jaundiced comment that "Few astronomers have studied comets from the observational standpoint more assiduously than did Crommelin, and if ever the futility of the direct observational approach to theoretical problems were clearly demonstrated the present suggestion remains as a monument" (*The Comets and Their Origin*, p. 156). He then goes on to refer to the "absurdity" of the conjecture and opines that it would not even be worth mentioning except that "the literature of astronomy shows that almost equally absurd notions are still actively entertained in regard to many problems." It is perhaps ironic that in Lyttleton's own speculation as to the nature and origin of comets, he admitted that he had no observational background in the subject and approached it from a theoretical perspective only. His own ideas have turned out to be as far from the mark as those of Crommelin. The lesson to be learned here is surely that neither observation nor theory alone and isolated from one another can hope to arrive at the real nature of a phenomenon.

Back to Crommelin however, this present writer does not see his conjecture as being as absurd as Lyttleton thought. It is certainly incorrect and would indeed be absurd if put forward today when knowledge of both solar prominences and comets has moved a long way since Crommelin's time. But, given the state of knowledge back then, the idea that an eruption of gas from the Sun might manage to condense into a swarm of particles (in agreement with the then-current comet model), whilst possibly a little farfetched, was hardly absurd.

Out of the Exploded Planet?

Chapter 2 addressed the controversial and, some might say, weird and wild hypothesis of Thomas Van Flandern. As was explained, this astronomer argued that for most of its existence, our Solar System had two major planets which do not exist today. In short, they blew themselves to smithereens quite recently on the cosmic calendar, the most recent exploding just over three million years ago. The reasons for presenting this scenario and the cause of these supposed explosions have already been discussed and we need not repeat the details here. For the present purposes, let's turn our attention to the theory of comets that Van Flandern argued followed from these planetary disasters.

Van Flandern proposed that the present population of comets, minor planets and meteoroids originated as debris from these planetary explosions, especially from the most recent one about three million years ago. Although the shattered planet hypothesis is not new, most of the earlier proponents did not see the catastrophe as having occurred so recently in the history of the planetary system. Yet, according to this astronomer, most of the observed features of the Solar System's population of small bodies can best be explained in terms of a cosmically recent event.

In order to demonstrate why this is so, Van Flandern asks us to consider the fate of debris ejected in every direction from the center of the planetary explosion. Some of the resulting fragments will go into elliptical orbits around the Sun and these orbits will differ considerably amongst themselves in terms of velocity and direction. Most of these orbiting fragments will, he argues, be eliminated eventually through the effects of gravitational perturbations by the major planets. Only two basic orbit types will remain after this gravitational sifting; approximately circular ones in which the fragment does not closely approach any of the major planets and ellipses that are so elongated that the orbiting fragment retreats to such vast distances from the Sun and planets that most of these have either not yet returned to the inner Solar System since the explosion or else are just returning for the first time in our own day or in the geologically recent past. The fragments that follow the almost circular orbits are asteroids whereas those in very elongated elliptical orbits are comets, as we have already discussed in Chap. 2.

For the present however, our interest is not primarily in his theory of the origin and dynamics of comets, but in his ideas as to what comets, these bodies thrown out of the exploding planet into wildly eccentric orbits, really are. We know that models of comets have changed over the years, from early notions that these are incandescent bodies whose appearance can bring heatwaves to Earth (one may note how many times descriptions of comets in nineteenth century records make mention the weather being hot at the time) to the widespread model of the earlier decades of last century which understood them to be flying gravel banks to the dirty snowball or icy conglomerate model put forward by Fred Whipple in 1950. The latter model—the diametric opposite of the one which understood them to be hot incandescent bodies—gained in popularity and was generally accepted within the astronomical community by the time the Giotto spacecraft revealed the solid nucleus of Halley's Comet in 1986. True, the nucleus was a lot darker than most had imagined it would be, but for the majority of astronomers that only meant that the surface of the snowball was a lot filthier than Whipple had originally thought, not that his basic model was flawed. Indeed, there had been strong hints for several years prior to the 1986 Halley apparition that comet nuclei were far from snow-white and pristine. Analysis of the reflectance spectra of relatively inactive comets far from the Sun indicated nucleus colors similar to those of the dark asteroids of the outer main belt; however this realization did not extend much beyond the small sect of professional cometary astronomers until it was dramatically confirmed by the Giotto images.

Whipple's model neatly explains what had hitherto been the most puzzling features of these bodies. As the icy object draws closer to the Sun, water ice and other frozen materials sublimate into the surrounding space. As the gases drift away from the nucleus, they carry off tiny particles of dust and the resulting dust-and-gas cloud swells to a truly enormous volume by comparison with the small dimensions of the solid nucleus. Amazing though it may seem, a nucleus just 3 or 4 miles in diameter becomes surrounded by a tenuous cloud of dust and gas over a hundred thousand miles across, from which may stretch a tail well in excess of a million miles long. The whole thing is made visible through a combination of sunlight reflected from, and scattered by, the particles of dust, plus fluorescence of the gases themselves; the comparative contributions of each varying from comet to comet and even in the same comet at different points in its orbit.

When this model was first postulated, the impression was gained that comets and asteroids are quite distinct in their physical nature. In short, comets are icy whereas asteroids are rocky. In more recent years, we have learned that these two classes of object are not as far apart in their nature as earlier thought. In fact, the distinction between them has become quite blurred. While some comets do appear to be very icy and fragile and many asteroids are rocky and apparently devoid of volatile substances, there is a wide intermediate range with objects that appear to be basically rocky and yet contain enough ice to drive at least a low level of cometary activity as well as objects that were once strongly active but have subsequently turned into dormant bodies that look very much like asteroids.

Van Flandern was ahead of his time when, in 1993, he stressed the similarities rather than the differences between these classes of minor Solar System object. He drew attention to speculation (later confirmed) that at least some of the asteroids that were not members of the main belt might be extinct or dormant comets as well as to the similarity in the composition of dust particles collected from the upper atmosphere during meteor showers associated with certain comets and the spectra of the majority of asteroids. That is to say, cometary dust particles are found to have broadly similar compositions to carbonaceous meteorites and most asteroids are dark C-type objects, suggestive of a similar carbon-rich composition. Then there are the bright fireball meteors which arrive on orbits stretching out beyond that of Jupiter, strongly suggesting association with comets, but which nevertheless have sufficient tensile strength to penetrate deeply into Earth's atmosphere. Finally, Van Flandern points to the detection by radar of the nucleus of Comet Encke. This yielded a radar reflectivity of around 10 % which is rather similar to that observed for asteroids that have been detected by radar and which implies that the surface material of Encke is, like that of the detected asteroids, non-porous and probably rocky.

At this point, Van Flandern appears to be moving in a similar direction to that taken by a few astronomers before him. For example, no less a cometary expert than Jan Oort, who in the early 1950s proposed a comet model in which their nuclei consisted of a single (or, at most, a very small number of) solid rocky monoliths similar to asteroidal bodies. Critics of this model argued that it made comets just too similar to asteroids and did not explain how the former could sprout diffuse heads and tails while the latter did not. Interestingly, one suggestion was that solar heating and rapid changes in temperature could cause the surfaces of such bodies to crumble and release dust. Although this process is inadequate to explain either the strength of cometary activity or the full range of forms that such activity takes (or to explain it at all at other than small solar distances) an example of just this sort of process has now been found! The rock comet Phaethon, parent body of the Geminid meteor stream, shows evidence of this type of activity when close to its very small perihelion distance and it is thought that this is how the small asteroid was able to build up its impressive meteor complex.

Nevertheless, Van Flandern points to other evidence that on the face of it appears to contradict this notion that comets are relatively rigid and strong rocky bodies. In particular, he raised the issue of the comparative ease with which many have been observed to split. These events happen too frequently to be explained by collisions with large meteoroids, yet they are hardly the sort of events that we should expect from strong rocky bodies!

The model that he suggests neatly explains these apparently contradictory properties of comets. Taking, as his starting point, the discovery of satellites in orbit around some asteroids, Van Flandern proposes that comets also are bodies accompanied by satellites. He defines satellites very liberally, suggesting that the secondary bodies around comet nuclei range in size from boulders to dust particles. Comets therefore "differ from minor planets in that small particles down to dust size could remain in orbit around the primary nucleus indefinitely, until the comet has been in close proximity to the Sun for hundreds or thousands of years." (DM, p. 196). Moreover, he does not abandon the dirty snowball model completely as, in addition to being surrounded by these vast numbers of satellite particles, the solid cometary nuclei may also contain "a full complement of frozen volatiles" which is only slowly depleted by their very infrequent visits to the solar neighborhood. By contrast, asteroids that remain constantly in the neighborhood of the Sun, which is also the relatively high density region of the planetary system, will long ago have become depleted of both frozen volatiles and most of their smaller satellite debris. In a sense, this model combines aspects of both Whipple's hypothesis and the older gravel bank conception of cometary nuclei, as well as echoing some features of the monolithic model of Oort and others.

The satellite model, in the opinion of its proposer, offers solutions to several types of observations that have otherwise proved to be difficult to explain. The first of these concern the reports (more frequent, it seems, in older literature where visual accounts were paramount) of a granulated aspect of the nuclei of certain comets. Although some of these appear, superficially, to relate to split nuclei, a study by Z. Sekanina shows that the details of most of these reported observations are not consistent with the dynamics of true nucleus schisms. Various possible explanations have been put forward for these reports. Photographic instances might have been caused by ghost images, plate flaws or guiding problems. Both photographic and visual observers may also have become confused by background stars (although these instances should be amenable to verification simply by consulting a good star atlas). There are also certain types of events within the comet itself that might to some degree resemble, but not be, true nucleus disruptions. Jets of gas and/or dust from an active spot on the nucleus surface may release a bright knot of material that could imitate a sub-nucleus. Sometimes, the apparent granular appearance may simply be in the eye of the beholder. In this respect, the writer notes that, when observing the coma (not the nuclear region) of Comet Holmes on one night in 2007, I had the strong impression that the entire head of the comet had an appearance which could indeed be called granular, except that the grains were extremely small. Although suspicious of psychological explanations for difficult-to-explain observations, I did actually wonder whether my impression was simply subjective and based on a subconscious association between the comet's appearance at that time and that of a ball of thistle-down. Be that as it may, it might be noteworthy that I was not the only observer to see this appearance, both during 2007 and during the comet's earlier mega-outburst in 1892. Van Flandern, in conformity with his hypothesis, interprets reports of granular nuclei as being genuine observations of the satellite system surrounding the central body. Maybe he would explain the appearance of the coma of Comet Holmes in terms of satellites escaping following the outbursts, although the mechanism by which this might happen is not at all obvious.

Secondly, Van Flandern proposes that the lightcurve of the nuclear region of Comet d'Arrest (the best observed and determined lightcurve for a comet nucleus at the time of his writing) betrayed a triple-maximum/triple-minimum form similar to that of some asteroids, for instance Betulia, for which an explanation invoking the presence of satellites had been proposed.

Comets can retain their satellites for a long time if they stay clear of large planetary bodies and, especially, of the Sun. A comet approaching the latter to within a distance of about 0.005 AU will have its sphere of influence, i.e. the distance from its center at which it can hold onto a satellite body, reduced to a value equivalent to its own radius. Any comet venturing that close to the Sun will therefore be stripped of its satellites and appear to break up into a string of sub-nuclei, each of which will then assume the role of a separate comet. Some comets, most notably those belonging to the Kreutz family of "sungrazers", have approached the Sun to that very small distance while others of this family have come almost as close. Two of these have definitely split while a third probably did. Moreover, it is virtually certain that the members of the Kreutz family are fragments of a progressively disrupting number of comets, all dating back to a single object that began the splitting process over two thousand years ago. However, according to the satellite model, the first sungrazing approach should have stripped away all the satellites, so these later schisms presumably resulted from some other process—the splitting of the solid nucleus itself perhaps?

It may seem quite amazing that a sungrazing comet can survive even one perihelion passage deep within our star's corona and the immediate reaction to the dirty snowball model is to think that the very fact that some do indeed survive cast doubts upon that theory. It is therefore a little ironic that their survival tends to support rather than undermine the icy model. Rapid evaporation of an icy body actually cools the underlying ice, for the same reason that pouring alcohol onto the skin feels cold. This loss of heat through evaporation slows the evaporative process and tends to prevent the comet from overheating as a purely rocky body might. Of course, much ice will inevitably be lost and small sungrazers certainly do evaporate away completely. Even large ones will not be capable of repeating too many sungrazing passages. This evaporative cooling phenomenon is acknowledged by Van Flandern who relies upon the presence of ice mixed with rock in his model to prevent catastrophic thermal stress in sungrazing comets.

The latest research on the sungrazing comets of the Kreutz family indicates that these ultimately sprang from a parent object that appeared around the year 214 BC and was probably identical with the one recorded by the Chinese that year. Unfortunately, little was said about this comet, except that it was bright and a broom star. One might think that "bright" would have been an understatement for the parent of the sungrazing group, but the paucity of information about this object suggests that it does not seem to have been exceptionally brilliant—just bright. Perhaps its surface was covered by a thick layer of refractory crust that not even a sungrazing encounter could completely purge. In any case, it seems that this object cracked or was in some way weakened as it passed perihelion, but held together for several hundred years before breaking into two large fragments far from the Sun. These fragments are thought to have arrived at their separate perihelia toward the end of the fifth century, one of them probably having been recorded as the magnificent object of February 467. Over the centuries, these fragments approached the Sun several times and progressively fragmented, evolving (or decaying, whichever way we look at the situation) into the very large family of today.

It is interesting to note that a possible Kreutz comet may also have appeared in the year 302. According to brief Oriental records, a comet was seen in daylight (apparently *only* in daylight!) sometime between mid-May and mid-June of that year. From Earth's perspective, any Kreutz comets appearing at that time of year come and go almost in the line of sight with the Sun and remain beyond it, making any sightings from Earth very difficult unless they t become bright enough to be seen close to the Sun in full daylight. The circumstances of the 302 comet therefore fit that of a Kreutz member, although this is by no means a foregone conclusion. If it was a Kreutz however, it probably started life as a fragment that split away from the 214 BC object close to perihelion, indicating that this object not only cracked but also lost some of its bulk at that time. That would not be surprising according to the standard comet model and further indicates that this object was becoming unstable due to the extreme stresses suffered during its periodic close encounters with the Sun.

As of mid 2015, nearly 3000 tiny comets, the vast majority being Kreutz members, have been found in data from the SOHO space-based solar observatory, but the last sungrazer having a moderately bright intrinsic magnitude was back in 1970. Nevertheless, though intrinsically rather faint, Comet Lovejoy of 2011 was well placed, at least for southern observers, and turned on a glorious display in the early morning skies around Christmastime that year. Figure 4.2 gives a good idea of the naked-eye appearance of this fine object, while the infrared image in Fig. 4.3 reveals the dusty composition of the comet's tail.

But returning to Van Flandern, what are we to say about his interesting hypothesis?

The idea that comets can have satellites is not in itself a particularly weird one. Many asteroid satellites are now known. Moreover, some of the objects in the Kuiper Belt, the home of most of the short-period comets, are also known to have satellites so there is nothing strange in thinking that similar bodies also orbit comet nuclei. Even the existence of myriads of dust-particle satellites surrounding some nuclei in a cloud does not seem too outland-

Fig. 4.2 Comet Lovejoy over Santiago, December 26, 2011 photographed by ESO Photo Ambassador, Yuri Beletsky (*Courtesy*: Y. Beletsky (LCO)/ ESO)

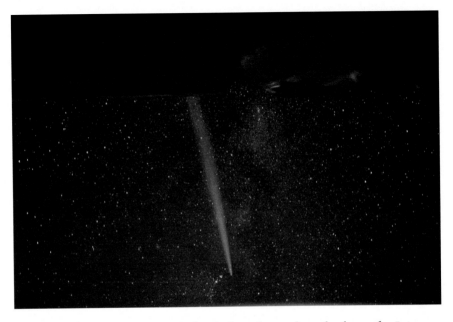

FIG. 4.3 Comet Lovejoy imaged at infrared wavelengths from the International Space Station, December 25, 2011 (*Courtesy*: NASA/crew of ISS)

ish a suggestion. However, since Van Flandern gave voice to his speculations on these matters, high quality observations of comets, including those made *in situ* by space probes, have provided a much clearer picture of the nature of these bodies and there is as yet no evidence of the existence of satellite bodies, at least none associated with the objects most thoroughly studied. The number of those well studied objects is, of course, only a tiny sample of the whole population and these negative observations do not rule out the presence of satellites orbiting some comets, but they do demonstrate that the typical range of cometary phenomena does not depend on these objects having satellites. The objects observed at close range by space probes were not peculiar in their behavior. On the contrary, they behaved just as comets are expected to behave, yet without the need for satellite bodies or clouds of orbiting dust grains. Some instances of apparent schisms amongst comets might be explicable in terms of escaping satellites and a few cases of double nuclei may be explicable on the satellite model, but from the consideration of the motion of split comets, especially within the context of the extensive research by Sekanina, these examples are very much in the minority, if indeed they exist at all.

Electrical Comets?

It is interesting and, we must confess, a little amusing at times to read through the brief notes on various phenomena recorded in science journals published during the closing decades of the nineteenth century and note how many times electrical phenomenon is given as the suggested explanation. Anything from unidentified sounds to luminous atmospheric events are presumed to be electrical phenomena as if this umbrella term is expected to eventually explain all remaining mysteries. This is a reminder to take heed of Fred Hoyle's comparison between the tendency of late nineteenth-century scientists to explain everything as electrical and what Hoyle saw as the tendency of recent years to explain away a wide range of astronomical phenomena in terms of black holes. As electricity was the in vogue explanation a little over a century ago, so black holes became the in explanation of more recent times.

Moving on from Hoyle's caution though, history shows that electricity eventually assumed its rightful place and, though important, was no longer seen as being the panacea for all mysteries. At least, not in the collective mind of mainstream science. But in science as in every intellectual tradition, main streams have many tributaries and some of these flow through weird and unconventional meanders while others are so wild as to miss the mainstream altogether and plunge out of sight as foaming cataracts navigable only by the most daring of independent thinkers.

Prominent amongst the latter was the strange and controversial figure of Immanuel Velikovsky who in 1950 dropped a literary bombshell in the form of his *Worlds in Collision*. Very briefly, this author speculated that during early historic times, a body of planetary dimensions was hurled out of Jupiter and swept through the inner Solar System as a sort of megacomet. This object made close approaches to Earth and Mars, causing great chaos on both planets, before eventually settling down into a nearly circular orbit intermediate between Mercury and Earth, where it remains today as the planet Venus. Velikovsky looked to ancient writings for evidence of his hypothesis as, remember, all this took place after Homo Sapiens had not only appeared on this planet but also developed civilization and culture. Alleged evidence came from many sources. The oldest Old Testament books of the Bible contained passages that appeared to describe phenomena that might be explained by very close encounters with a planet-sized comet. Ancient writings of other cultures likewise contained accounts of phenomena that appeared to fit Velikovsky's ideas. In particular, he found curious references in Mexican accounts of Venus "smoking" in ancient times. What are we to make of this "smoking"? Although some type of atmospheric distortion has been suggested, Velikovsky drew attention to the Mexican description of comets as "smoking stars" and concluded that if Venus was said to smoke, then this is tantamount to saying that it was originally a comet.

Velikovsky was well aware that Venus was far too massive a body, with far too strong a gravitational field, to allow the escape of the large amounts of gas and dust needed for it to smoke in the manner of a comet. At least, he knew that this could not happen if gravity was the only force to be considered. If his model was to work at all, something other than gravity must have played a vital role in the cometary activity of Venus … and this "something", he believed, was electrical repulsion. Once again, he turned to ancient tales and records and was able to convince himself that our ancestors saw things in the sky that we do not find there today but which appeared, from their descriptions, to be extremely intense electrical discharges, even including electrical arcs between planets moving on unstable orbits. (How an electrical arc can become visible in the vacuum of space is something that we might ponder!) Not only did he speculate that cometary activity associated with Venus might be explained in terms of electricity, but he also came to question the role that gravity played in holding the planets in orbit around the Sun. He argued that the Sun possessed an electrical charge capable of keeping the Solar System in order, thereby relegating gravity to a far less central role in celestial mechanics.

Needless to say, the publication of *Worlds in Collision* created quite a stir, to put it mildly. Most scientists dismissed it outright and, by and large, the Velicovskian enterprise was deemed not just weird but seriously crackpot. Yet, he did have his supporters, albeit mainly beyond the fringes of orthodoxy.

Thus, 10 years or thereabouts after the publication of Velikovsky's book, an engineer from Flagstaff, Arizona, by the name of Ralph Juergens began to collaborate with him, relating especially to the historical evidence of electric phenomena in the heavens. Through this collaboration, Juergens came to conclude that the Sun is the most positively charged object in the Solar System and that it is effectively the center of the electrical system which constitutes its family of planets and other bodies. In his view, all of the Solar System bodies carry electrical charges "but the Sun itself is the focus of a cosmic electrical discharge" and that it is this "cosmic electrical discharge" that probably constitutes the source of its radiant energy (*Immanuel Velikovsky Reconsidered*, p. 6). Apparently, the Sun shines by electrical and not thermonuclear processes. If that were true, presumably all stars are foci of cosmic electrical discharges and modern astrophysics is seriously in error.

According to the Juergens model, the most negative regions will be those parts of the Solar System that lie furthest from the Sun. This, he believes with some justification, is where comets spend most of their time. Based upon this, he argues that the nucleus of a comet will come to acquire the strong negative charge of its environment and that this will lead to electrical stresses as it approaches the strongly positively-charged Sun. He argues that in this situation "A space-charge sheath will begin to form to shield the interplanetary plasma from the comet's alien field" and that it is this sheath that develops into the cometary coma and tail as the object draws closer to the Sun.

Although this comet model is far from the one most widely accepted today, it did win a few supporters within the physics community, although the two most prominent "converts" to the model diverged in opposite directions over the finer details.

Staying closest to Juergen was Australian physicist Wallace Thornhill. According to Thornhill, as a comet approaches the Sun, electrons will be stripped from the surface of its nucleus, resulting in "a huge visible glow discharge" which he equates with the coma. However, as the comet draws ever closer to the Sun, the discharge switches to the arc mode. This change in discharge mode results in a number of bright cathode spots of high current density on the surface of the nucleus and these will be sufficiently energetic to etch out circular craters, as well as burning the surface black. Each of these arcs will form a cathode jet that "electrically accelerates the excavated and vaporized material into space."

This model nicely accounts for the very dark surfaces of comet nuclei as well as broadly agreeing with the "standard" icy model in so far as it understands active comets to be objects undergoing a process of slow dissolution. Nevertheless, a very different conclusion was reached, albeit from the common starting point of the Juergen model, by Cornwell University physics and mathematics lecturer James McCanney. According to this author, "a comet involved in the discharge of the solar capacitor will continue to grow in size and mass" (*The Nature of and Origins of Comets and the Evolution of Celestial Bodies-1*, **Kronos**, 9, 1, Fall 1983). Contrary to what nearly everyone else who has studied the phenomena believe, McCanney argues that at least some comet tails—principally the curved appendages of dust such as displayed by comets Donati, Bennett and the like—actually represent matter from the Zodiacal disc falling into the comet and building up on its nucleus. Comets do not slowly dissolve into the Zodiacal cloud as the majority of astronomers believe but, because of their electrical attraction, actually accrete particles from the cloud and build up their mass, eventually, in McCanney's opinion, evolving into planets. In 1984, McCanney proposed a definitive test of this hypothesis in the form of observations of tail material directly moving toward, rather than away from, the nucleus. Needless to say, all observations thus far have indicated movement in the opposite direction (Fig. 4.4).

What can we say about the electrical comet hypothesis?

First of all, we must admit that the wider theory on which it is based has little observational support, to put it charitably. Astrophysics has made great advances in the understanding of stellar processes and of stellar evolution through the application of the theory of nucleosynthesis. Not only does thermonuclear fusion explain how and why stars shine as well as explaining the various stages through which they pass during their main-sequence career and beyond, but it also nicely explains the synthesis of the range of elements which constitute the universe around us, including those making up our own bodies. No major role remains for electrical phenomena in the explanation of the energy, life, evolution and

FIG. 4.4 Comet Donati showing curved dust tail, October 5, 1858 (From *Bilderatlas der Sternwelt*, Edmund Weiss, 1892. *Courtesy*: Wikimedia)

death of stars and an electrical account of stellar energy leaves no place for the synthesis of the observed range of elements.

Nevertheless, it is conceivable that some form of the electrical hypothesis, as it is applied to comets, might still be true even though the wider thesis on which it was originally based is not. Can we concede that much to the model?

Little can be conceded to the McCanney version. We might see some small similarity between McCanney's hypothesis that comets accrete matter through their dust tails and Alfven's suggestion that they may form within meteor streams, but the evidence for McCanney is even weaker than anything that might be brought forth in favor of Alfven's suggestion. The structure of cometary dust tails is now regularly predicted according to the mechanical theory first proposed by F. Bessel and later extended by Th. Bredichin and, still more recently, by Z. Sekanina. Back in 1973, the latter successfully predicted in advance the extent of the dust tail of Comet Kohoutek by computing the trajectories of particles of varying sizes expelled from the comet's nucleus as they moved away from the comet under the opposing influences of solar gravity on one hand and solar radiation pressure on the other. Very briefly, the smaller particles are accelerated faster in the anti-solar direction by the pressure of sunlight while somewhat larger ones are accelerated less and influenced to a greater degree by the attraction of the Sun's gravity. The largest particles are only minimally affected by the pressure of sunlight and remain in orbit around the Sun close to the comet itself. By plotting the shape of the various curves followed by these particles, also taking into account the times at which they left the nucleus as well as the angle at which they are viewed from Earth, the shape and extent of a dust tail can be predicted for a variety of times and the presence or absence of an anti-tail (an apparently sunward extension) at any given date forecast with some confidence. Since 1974, these predictions have been made for many comets and it is no exaggeration to say that, thanks to this theory, these aspects of dust tails are more predictable than the brightness of comets, about which there is still much uncertainty. The very success of these predictions must surely count as sufficient proof that the accepted theory of these features is correct and that McCanney's alternative fails completely. Likewise, McCanney's suggestion that comets evolve into planets has absolutely no support in observation. Where, for instance, are the planets that would have evolved from comets of the Halley orbital type, according to his hypothesis? Not one planet has been observed in a Halley-type orbit, for which we can be thankful. On the other hand, there is plenty of support for the counter opinion that comets degrade over time. Not only do we see this happening in the formation of cometary tails and meteor streams, but a relatively large number of comets have actually been seen to fall apart before the eyes of observers.

Turning to the electric-comet idea in general, it must be said that it nowhere accounts for observed phenomena that are not better explained in terms of the icy model. Spectra of developing cometary coma are just as expected for an object composed of various frozen substances partially sublimating as it draws closer to the Sun. On the contrary, there is no spectroscopic evidence for the existence of the type of electrical discharge suggested by the electric hypothesis.

Electrical activity is not required to explain the black and cratered surfaces of comet nuclei either. Like any object exposed to the dangers of outer space, comets are vulnerable to meteoroid strikes and evaporating pockets of more volatile ices will also give rise to vents and craters, so there should be no surprise that comet nuclei have their pits and blemishes. As for their pitch-black surfaces, although this came as a surprise to most people, there is really nothing strange about it. Talk of ice and snow gives a misleading notion of what comet nuclei are really like. The ice of a comet is certainly not the sort of stuff that one would add to a cocktail. Mixed with the water ice and frozen gases is a good deal of organic material, some of which is very dark. Even a chunk of fresh cometary ice is likely to be as black as jet and as the ice at the surface of a nucleus boils off into space, much of this black organic material is left behind, building up into a solid crust having very low reflectivity. There is no need to burn a cometary nucleus with electrical discharges to turn it black. It contains more than enough dark material to do that job quite adequately.

So can the electrical model of cometary activity be dismissed entirely? Perhaps not. Someone once said that theological heresies are not so much total errors as doctrines that take a truth too far. In many instances, something similar could be said about scientific heresies. Even the weirdest idea can possess a grain of truth, however small that grain may be, and before one throws away the bathwater, it is wise to make sure that a tiny baby is not also hidden in there somewhere.

The existence of one unexpected electrical phenomenon associated with Solar System objects came to light through a series of curious and seemingly impossible observations, namely, twilight phenomena on the Moon. First hinted at in images beamed back by the Surveyor spacecraft, these phenomena were subsequently confirmed by astronauts on board Apollos 8, 10, 12 and 17. What the astronauts described was very similar to the crepuscular rays frequently observed on Earth following sunset or preceding sunrise. On our planet, these are easily explained as strips of light sky appearing between the shadows of distant geological features of even of clouds. Clouds so distant as to be below the local horizon can give rise to impressive sets of crepuscular rays fanning out from the sunset point on a clear horizon. But a shadow cannot be visible in a vacuum. Crepuscular rays are only seen in Earth's twilight because our planet is enveloped in a gaseous atmosphere in which solid particles are suspended. If the same thing occurs on the Moon, one might look for a similar explanation, except that we have long known that the Moon has no atmosphere and should therefore not give rise to any form of twilight phenomenon. What the Apollo astronauts saw should therefore be impossible. What they reported was a phenomenon that could not exist according to the accepted knowledge of the time—but they saw it anyway.

Fortunately, Apollo 17 left behind it an instrument known as LEM or the Lunar Ejecta and Meteorites detector. Its purpose was to judge the frequency of meteoroid impacts on the lunar surface by monitoring the dust kicked up by these events; however it unexpectedly detected something that nobody had expected but which finally solved the problem of the lunar twilights. Surprisingly, LEM detected a large number of moving particles around the time of each lunar sunrise. These particles moved more slowly than those raised by meteoroid impacts and they all appeared to be travelling in the same direction, from east to west across the terminator; the line of demarcation between lunar day and lunar night. It is as if the terminator is accompanied by a sort of perpetual lunar dust storm.

The explanation for this phenomenon, according to NASA's Timothy Stubs, is electrostatic repulsion. Stubs reasoned that if the day side of the Moon becomes positively charged and the night side negatively charged, horizontal fields at the terminator will push electrostatically charged dust particles sideways across this zone of demarcation. In that way, the terminator will always be accompanied by a moving line of elevated dust which, even in the absence of an atmosphere, will nevertheless be of sufficient height and density to give rise to a degree of twilight phenomena (Figs. 4.5 and 4.6).

If this phenomenon can occur on the Moon, it seems reasonable to suggest that it should occur on other Solar System bodies as well, including asteroids and maybe even comets. Indeed, David Jewitt has suggested this as one of the possible mechanisms giving rise to the activity in so-called active asteroids or main-belt com-

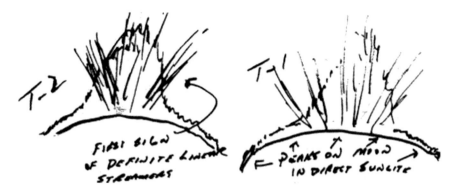

FIG. 4.5 Twilight phenomena on Moon—1 (*Courtesy*: NASA)

FIG. 4.6 Twilight phenomena on Moon—2 (*Courtesy*: NASA)

ets, although he does not believe that it is a major factor in these. Although the gravity of the Moon is too strong to allow the elevated particles to be swept away by the solar wind, such is not the case on very small asteroidal bodies. If a sufficient amount of dust is swept away through this process, the resulting dust tail might be observable from Earth and described as weak cometary activity.

One such asteroid which may show this effect is 3200 Phaethon, the parent body of the December Geminid meteor shower. Initially, this object was thought to be the rocky core of the nucleus of a very large comet which had long since lost its volatiles and ceased to be active. The main reason for thinking this was the presence of the strong Geminid meteor shower and the prevailing belief that only a comet could give rise to such strong meteor activity. Nevertheless, there was always a difficulty in understanding how a comet (presumably originating in the Kuiper Belt and migrating inwards over time) could end up in such a small ellipse with an aphelion distance well removed from the gravitational influence of a major planet. Moreover, not all astronomers were happy with the idea that comet nuclei possessed cores of solid rock, although some of the skeptics would probably have conceded that unusually large ones may contain such features.

It now seems more probable that Phaethon is truly asteroidal. Indeed, it appears to have originated as a fragment of the large asteroid 2 Pallas, one of the Big Four asteroids whose discovery led to the recognition of the very existence of this class of object.

Maybe there was some ice within Phaethon initially (as there may be in Pallas itself) but its small perihelion distance and very short-period orbit means that it cannot now cool down completely and the temperature even at its core remains too high for ice to be stable. Yet, the asteroid is active to some degree, as is evidenced both by the presence of the Geminid meteors and by weak activity sometimes observed when it is near perihelion as, for instance, on June 20, 2009 when a definite brightening and release of dust was recorded. According to Jewitt, the principal process driving this activity is thermal stress causing breaking of hydrated minerals on the surface due to the 1000° heating around the time of perihelion. He classifies Phaethon as a rock comet, essentially a non-icy asteroidal body gradually falling to pieces due to frequent thermal stress. Nevertheless, while granting that this process is the main cause of activity, it still seems possible that (given the asteroid's proximity to the Sun at perihelion) electrostatic repulsion might occur and assist in elevating the dust, released through the thermal stress, to levels where it can be more readily swept away by the solar wind. Larger particles of the type expected to end up in the Geminid stream may not be affected by this process, but it seems reasonable to think that finer particles could be. To this small degree therefore, some life may yet remain in the electric comet model!

Before leaving the field of electrical astronomy as we might call it, mention should also be made of an ingenious model put forward by physicist Charles Bruce (1902–1979). Although not concerning the electrical model of comets per se, Bruce, like Juergens, McCanney and the others mentioned earlier in this section, saw electricity as playing a far greater role in astrophysics and even in cosmology than mainstream science allows. Although accepting the orthodox thermonuclear model of stars, he argued that many types of stellar phenomena could be explained in terms of lightning discharges within stellar atmospheres. For example, he saw in the phenomenon of solar prominences examples of electrical discharges within the atmosphere of the Sun. Looking further afield, he explained the sudden brightening of stars in nova eruptions as being due to immense stellar lightning flashes within their atmospheres and he even predicted that super-sized lightning bolts should take place in galactic atmospheres and on this basis, successfully predicted the occurrence of extremely bright objects—star-like in appearance because of their distance—within distant galaxies. These galactic lightning bolts, as envisioned by Bruce, were remarkably similar to the quasars later discovered, initially by radio, and then by optical, astronomers. But Bruce went even further and postulated the existence of lightning bolts of truly cosmic proportions; 600 billion miles wide, 60,000 billion miles long and possessing temperatures of 500 million degrees Celsius. These mega-bolts, generated in cosmic dust distributed throughout the universe, were in Bruce's opinion, responsible for bringing the stars and galaxies into being. In a 1960 paper entitled *An All-Electric Universe*, he wrote:

In our atmosphere the [lightning] discharges may last for seconds, in stellar atmospheres for years, and in galactic atmospheres for tens or hundreds of millions of years, while on a still grander scale they may originally have enveloped the whole universe as we know it … electrical discharges have gradually condensed the matter from the primordial gas and dust of a general universal atmosphere, first into the galaxies, then from the condensed matter of the galaxies into stars. Discharges in the extended atmospheres of stars further condense the matter ultimately to allow of the formation of planets and satellites.

There are clear similarities between Bruce's hypothesis and those of Juergens, McCanney and company, as well as, in certain respects, to that of Alfven. Nevertheless, Bruce appears to have arrived at his ideas quite independently of these other folk and developed these astrophysical speculations from his expertise in the study of electrical phenomena in general.

Continuing astrophysical and cosmological research in the years since Bruce developed his hypothesis has left little room for cosmic electricity on the scale that he envisioned. Nevertheless, his opinion that lightning can be generated in clouds of cosmic dust might not be too far from the truth, albeit not on the grand scale about which he speculated. The presence in typical stony meteorites of chondrules—small inclusions that look like solidified droplets of once-molten rocky particles—necessitates the occurrence of very brief energetic events in the early Solar System. Indeed, because chondrules rate amongst the most primitive objects in the System, these energetic events (whatever they may have been) must have taken place at a very early date. Although several mechanisms have been proposed, the suggestion of nebular lightning put forward by, e.g., Fred Whipple is an interesting one which nicely accounts for the presence of these tiny objects. Lightning bolts generated within the dust of the early solar nebula would have briefly melted dust particles in their paths, thereby creating objects that would look very much like the chondrules actually found in chondrite meteorites. If nebular lightning flashes truly were the brief energetic events responsible for chondrules, the smaller types of cosmic electrical discharge spoken about by Bruce may indeed have played a role in our Solar System, albeit a far smaller one than Bruce himself imagined.

Antimatter Comets

If the hypothesis of cometary activity driven by electricity seems a little weird, what can be said about the suggestion that at least some comets are composed of antimatter?

The existence of antimatter poses a problem for astronomy. As mentioned earlier in this book, whenever subatomic particles are created in particle accelerators, they always appear in pairs; one particle of what we like to call ordinary matter and one of antimatter. The latter is a sort of mirror image of the former—a particle just like that of ordinary matter but having the opposite charge. The antimatter equivalent of an electron, for example, is the positron; a positive electron so to speak. However, mysteriously, the observable universe appears to be almost completely composed of ordinary matter. Observations at gamma-ray wavelengths of the region close to the core of the Milky Way detects the presence of some antimatter in that region, thought to be the product of certain energetic processes such as the radiation from hard X-ray binary stars. These binaries consist of relatively normal stars paired with collapsed objects such as neutron stars or black holes, from which matter is drawn from the companion star and accelerated to extreme velocities as it reaches the collapsed object. The energies involved are large, but comparatively local. The creation of antimatter near the center of the Galaxy does not solve, or even address, the problem of the missing antimatter on a cosmic scale.

Some cosmologists have suggested that there may be matter and antimatter galaxies distributed in equal measure throughout the universe. This, as we have already seen, formed an essential part of Alfven's cosmological model, although in his hypothesis, the antimatter region of the cosmos lay beyond the bounds of our observable universe. Yet, even if antimatter galaxies existed closer than Alfven thought, they would presumably not clearly distinguish themselves in our telescopes, so at one level this idea appears to offer a reasonable solution to the problem. Nevertheless, from our experience with particle accelerators, it is clear that particles of matter and those of antimatter are generated together, at the same place, so that any theory seeking to balance the matter/

antimatter budget by collecting both sets of particles in separate galaxies must find some way of segregating them into cosmologically different camps. This is difficult, to say the least. On the other hand, if the two types of particle remain mixed together, it is equally hard to see how our universe could continue to exist. When a particle meets an antiparticle, the two annihilate one another and emit a burst of energy in the form of gamma rays. That is how gamma-ray observations of the central region of the Milky Way discovered the presence of antimatter there, but it does not hold out good prospects for a universe in which the two are mixed together.

Nevertheless, the thought that some objects comprised of antimatter may manage to wander around the Galaxy has been entertained over the years by various folk and a number of suggestions have been made that implicate these hypothetical bodies as the cause of certain astronomical mysteries.

One of the earliest of the hypotheses of this type was proposed in the 1940s by physicist Vladimir Rojansky in a paper entitled *The Hypothesis of the Existence of Contraterrene Matter*. In this paper, Rojansky proposed that some comets and meteoroids within the Solar System might actually be lumps of antimatter, or contraterrene matter as he preferred to call it. Suppose a relatively large lump of this stuff orbited the Sun in an elongated ellipse. As it drifted deeper into the planetary system, atoms of ordinary (or terrene to use Rojansky's terminology) matter from the solar wind, together with particles of cosmic dust, would inevitably impinge upon its surface and annihilate. The results of such a process might be observed as cometary activity.

Rojansky did not propose that all comets are made of antimatter, but that antimatter bodies moving through the Solar System might imitate the behavior of ordinary comets and could be mistaken for such. Yet, whilst we might agree that his suggestion is theoretically possible, no evidence suggests that any object of this nature has found its way into our catalogues of comets and the thought that such sizable lumps of antimatter could survive for long in our overwhelmingly terrene galaxy is problematic. Their continued existence within the Solar System would be especially precarious, given the presence of cosmic dust, micrometeoroids and the solar wind. According to calculations by Canadian astronomer

Martin Beech meteoroid-sized antimatter objects could not survive in the Solar System environment for more than 10,000–100,000 years and that any such objects must either be new arrivals or hyperbolic bodies passing through the Solar System from places far afield.

Nevertheless, Rojansky was not alone in postulating the existence of antimatter bodies. In 1947, Osmania University's M. Khan proposed that the mysterious glassy tektites found in various regions of Earth may have resulted from the explosive annihilation of antimatter meteorites. This suggested explanation for these puzzling objects did not, however, gain much support from the wider science community.

A speculation that gained somewhat wider publication, if not necessarily wider acceptance, involved that notable event that seems to have been the focus of all manner of unconventional hypotheses, the Tunguska fireball of 1908. In 1958, P.J. Wyatt of Florida State University proposed that this was actually an antimatter meteorite which annihilated in the lower atmosphere. The reason for entertaining such a wild sounding notion was, essentially, the apparent lack of any meteoritic material found at the site of the event. Recent studies, as we shall see later in this chapter, have shown that this event was actually a lot less mysterious than formerly thought. In particular, the detailed study by Z. Sekanina found that the description of the fireball, the altitude at which it exploded and its most probable orbit were all entirely consistent with the body having been a perfectly normal, small, stony Apollo asteroid—composed of perfectly normal matter. The Tunguska event was intrinsically no more mysterious than any other bolide. Only larger than most. Fortunately for us, bolides as large as Tunguska are rare.

Nevertheless, during the middle years of last century, the Tunguska event still seemed very mysterious to many people and Wyatt's ideas struck a chord with some. W. Libby and C. Cowan, for example, noted the presence of unusually high levels of carbon-14 in tree rings formed in 1909 and suggested that this may have been associated with the previous year's event, implying that the explosion may have been nuclear. Other alleged evidence for its nuclear nature proposed during the 1960s was the claim that radioactive material had been found in the soil throughout the region of the explosion. However, this was later shown to have been due to Soviet atomic bomb tests during the 1950s, not to the Tunguska explosion. No evidence for radiation accompanying the latter has ever been found.

A serious problem with the antimatter-meteor hypothesis is the depth to which the Tunguska body managed to penetrate into the atmosphere before exploding. That is also, by the way, an objection to the other popular hypothesis that the body was a small comet rather than an asteroid. An antimatter body would have encountered enough matter at far higher altitudes for the annihilation process to have taken place earlier and at a greater height above the ground.

Perhaps the weirdest of all the antimatter hypotheses was the one put forward by D. Ashby of Culham Laboratory together with C. Whitehead of the U.K. Atomic Energy Research Establishment in 1971. By monitoring the sky with gamma-ray detection apparatus, these scientists found unusually high numbers of gamma rays, having an energy level characteristic of electron/positron annihilations, within the atmosphere. Now, positrons can be generated indirectly by thunderstorms as lightning creates unstable isotopes of nitrogen-15 and oxygen-15 which subsequently decay. However, as there were no thunderstorms present at the times of the Ashby and Whitehead observations, they proposed that the observed gamma rays resulted from small antimatter meteoroids (perhaps from antimatter comets?) entering Earth's atmosphere.

Ashby and Whitehead did not merely postulate the arrival of small antimatter meteoroids into Earth's upper atmosphere however. They proposed a way in which micrometeorites composed of antimatter could survive their downward drift through the air until they reached the surface of our planet—where they are sometimes observed as ball lightning! The relatively long life of these tiny bodies within our atmosphere was supposed to result from a potential barrier that may form between a speck of antimatter and the ordinary matter surrounding it. Ashby and Whitehead suggested that, as the antimatter micrometeoroid floated down through the air, atmospheric molecules would not always carry enough energy to penetrate this hypothetical barrier and annihilate the antimatter particle. The anti-atoms at the surface of the antimatter particles would instead become anti-anions as their positrons become stripped away by the photoelectric effect, just as ordinary atoms can be transformed into cations by the stripping away of their electrons.

This proposed explanation for the phenomenon of ball lightning never gained wide acceptance. Besides the broader problems associated with antimatter particles of micrometeoroid size and larger within the Solar System, the hypothesis stalls at the completely hypothetical and unsubstantiated postulation of a barrier between objects comprised of the different forms of matter. The only possible supports for the hypothesis are the rare accounts of ball lightning that could be interpreted as revealing evidence of radioactivity. Avid collector of nature's anomalies, William R. Corliss, lists three cases in his book *Lightning, Auroras, Nocturnal Lights and Related Luminous Phenomena* which may have involved radiation injury, although Corliss also points out that the effects noted could just as easily have been "induced by electric shock, perhaps amplified by the traumatic experience of being close to ball lightning" (p. 85).

One incident dating from November 24, 1886, involved the appearance of a ball of light within a house. Members of the family who witnessed the event were immediately affected by "violent vomiting" and the appearance of swellings on parts of their bodies: swellings which had turned into large marks by the following day and, on the ninth day, peeled to reveal "virulent large blotches." There was also hair loss on the side exposed to the ball lightning. Another curious feature noted was the withering of trees near the house, also only apparent after the ninth day.

The other two instances noted by Corliss involved a momentarily blinding, followed by head and neck pains, of a ball-lightning witness on July 2, 1893 and memory loss for several hours following an electric shock induced by another encounter with ball lightning on an unspecified date and location. The after effects recorded in association with these latter two instances were almost certainly caused by electric shock. The former may also have been simply electrical, possibly exacerbated by psychological trauma (the witnesses were said to have been very frightened and to have thought that the end of the world was upon them). In any case, these reports constitute a very small minority of cases. The majority of encounters with ball lightning do not include after-effects suggestive of exposure to radiation. The writer is personally aware of someone who was actually singed by ball lightning but suffered no ill effects of the type suggesting exposure to radiation. Certainly, there are still mysteries surrounding the subject of ball lightning and there may be some surprises as further knowledge comes to light. Antimatter micrometeoroids, however, are unlikely to be amongst them.

Finally, the proposed existence of antimatter comets was seen as a possible explanation for the mysterious phenomena of gamma-ray bursts. A suggestion put forward in the final decade of the former century proposed that these brief and mysterious bursts might be due to collisions between very small antimatter comets and equally small objects comprised of ordinary matter. The burst of gamma rays as the comets mutually annihilated would be brief but readily observable from the gamma-ray telescopes placed in Earth orbit.

This hypothesis is ingenious, however, quite apart from the general lack of evidence that antimatter comets exist (indeed, by what must now be considered strong evidence that they do not) the frequency of these collisions necessarily imply a stupendous number of small comets (both ordinary-matter and antimatter ones) existing within the Oort Cloud.

Moreover, as further observations of gamma-ray bursts were collected, it became clear that these events lie far beyond the Oort Cloud. Indeed, they lie far beyond our galaxy. In 2002, a supernova in another galaxy was found to be associated with a gamma-ray burst and other supernova/gamma-ray burst associations have been discovered during subsequent years. Dying stars, not colliding comets, are now recognized as the sources of these violent and energetic events.

Cosmic Serpents and Cosmic Winters

In the previous section, reference was made to the wild and weird speculations of Immanuel Velikovsky. So unconventional, so way out, were his hypotheses that there was a time when even the mention of his name in main-stream astronomical circles was enough to have astronomers throwing up their hands in horror.

Part of the revulsion aroused by Velikovsky's ideas was due to his eccentric account of the origin of the planet Venus, but there is good reason to think that part of the problem also lay with his method. Whatever we might think of the actual details of his hypothesis (and these details are surely far removed from the real universe) his method stands or falls on its own merit. The starting point of his hypothesis lay with the conviction that ancient stories more or less faithfully recounted events of an astronomical nature. Furthermore, these events were not necessarily simply equated with phenomena known and recorded in more recent times. The ancient peoples saw things that have not been observed for centuries, possibly millennia, and it is a valid project for the modern astronomer, in the opinion of Velikovsky and his supporters, to interpret the nature of the phenomena that gave rise to the accounts left (often framed in the language of mythology) by our distant forefathers. Although we must admit that Velikovsky's interpretation went too far into the left field, that does not necessarily discredit his approach per se.

There is, let it be said, no *a priori* reason for thinking that all astronomical events, or any kind of natural phenomenon for that matter, should occur on a time scale that is more or less commensurate with the average lifetime of a human being or even with the span of recorded human history. Perhaps there are phenomena that occur only once in a thousand, or ten thousand, or a million years. We may read an ancient account of (say) fire in the sky and immediately rationalize it as a bright aurora. Chances are we would be right, but there is also a possibility that what was seen was something so rare that it only happened once during the entire span of human history and may not occur again for another hundred thousand years. A television interviewer once asked Michael Palin if he believed that there still remained undiscovered places on Earth. With typical Palin humor, he replied that he did believe this but "the trouble is, no-one knows where they are". We might say much the same about undiscovered astronomical phenomena visible from Earth. Yes, there may well be such things, but we don't know what they are. In this context, it might be worth recalling that the electrophonic sounds generated by meteorites were dismissed as being psychological (a euphemistic way of telling witnesses that they imagined it) until the work of Professor

Colin Keay in the 1970s and lightning sprites were not officially discovered until the late 1980s, even though they can be visible with the naked eye. I wonder how many witnesses of this phenomenon had dismissed it—or had it dismissed for them—as after-images left by flashes of lightning?

As we saw earlier, Velikovsky inspired certain independent thinkers to elaborate his views on the role of electricity in cosmic events, following from his speculation that some of the events recorded in ancient times related to the observation of cosmic discharges of a type not occurring today. These thinkers were, nevertheless, still outside the boundaries of mainstream thought. However, in the latter decades of last century, the general opinion concerning cosmic collision, if not cosmic electrical discharges, made it more "respectable" for a few daring astronomers to speculate about such matters and even to claim Velikovsky as part of their inspiration, whilst remaining within the broader main stream of astronomical thought.

The two most prominent of these astronomers are Victor Clube of the University of Oxford and Bill Napier of the Royal Observatory, Edinburgh. These astronomers, while rejecting most of the specifics of Velikovsky's model, nevertheless agreed with his approach in general and looked to ancient writings for evidence of astronomical events that only exist today as faint shadows of their former selves.

In essence, Clube and Napier argued that in prehistoric times (possibly as long ago as the late Stone Age) a very large short-period comet—the predecessor of today's Comet Encke and its extensive Taurid meteor stream—was an almost constant feature of the night sky. So huge did this appear, that early humanity saw it as a god; indeed, as the High God of the cosmos. Over the centuries, this comet faded and broke into a number of smaller ones, each of which eventually faded to obscurity, leaving behind only ancient tales of a god who was eventually replaced by many gods and finally departed from human vision (Fig. 4.7).

Both the original comet god and the successor gods could interact with the human race. We know that as comets age, many particles are spread along their orbits and will enter Earth's atmosphere as meteors if our planet approaches sufficiently closely to the cometary orbit itself. This original giant comet, in the opinion

FIG. 4.7 Comet Encke from Mercury, as imaged by MESSENGER space-craft November 11, 2013 (*Courtesy*: NASA/JHUAP/Carnegie Institute of Washington)

of Clube and Napier, not only left in its wake "regular" meteoroids like the familiar Taurids of our own day and age, but also a large number of sizable chunks of rock ranging in size up to over 300 feet in diameter. In those distant times, the Taurids were both more numerous and a good deal more dangerous than the weak (albeit long-lasting) drizzle of shooting stars and fireballs that we observe nowadays. Frequently, as Earth crossed the orbit of their parent comet, some of these large objects entered our atmosphere and exploded with a force such as that observed in the Tunguska region of Siberia in 1908. Because of the thunderous roar generated by these events, they were understood as the thunderbolts hurled by the gods. Tales of Zeuz (the Greek High God of pre-Homeric times, later reduced to the head of a pantheon of lesser deities) hurling thunderbolts which wiped out whole cities and all of their inhabitants in one fell swoop do indeed sound more like accounts of Tunguska-like bolides than the strokes of lightning which we call thunderbolts these days.

Moreover, even larger bodies are supposed, by these authors, to lurk within the Taurid stream. An encounter between the Earth and one of these objects of asteroidal dimension would be catastrophic indeed. An explosion of multi-megaton force would not only devastate a large area of our planet's surface, but even more seriously, would waft such large volumes of dust into the stratosphere as to seriously diminish the degree of the Sun's light and heat reaching the surface of our world. In the stable stratosphere, dust particles could remain suspended for many months at least, plunging the Earth into the cosmic winter for which one of the Clube and Napier books was entitled. The resulting famine would surely be interpreted as punishment by the comet/god according to this hypothesis.

This regression in theological belief from a High God to a pantheon of lesser gods—from monotheism to polytheism—goes against a long tradition of anthropological accounts of the evolution of religious belief in which a primitive animism evolves into polytheism and eventually into monotheism as one of the gods gained mastery over the rest and finally becomes the only god, the other supernatural entities being relegated to the role of angels, demons and such like. Nevertheless, anthropological field research as distinct from armchair hypotheses indicates that quite the opposite happened in the real world. There is evidence that the most ancient form of religious belief involved the worship of a High God who was in some manner associated with the sky or manifested through the sky. Apparently this original monotheistic belief waned over time and the High God's power and authority was, in the belief systems of later generations, progressively taken over by supernatural beings who were originally thought of as being less than divine, but who subsequently became elevated to positions of (if we might use the term) secondary divinity. These gods had power, not over the whole of reality, but over certain aspects of life or nature. They were, in fact, not unlike the saints of the mediaeval church. Indeed, one might easily see the historic elevation of the saints as another example of the decay of an earlier monotheism into polytheism. Apostolic Christianity was vital, charismatic and prophetic and its adherents had a very real awareness of the presence of God. But as it became more formal and institutionalized, this early spiritual vitality was sapped and God was perceived as being increasingly remote. Hence the belief in intermediaries whose power, though infinitely less than that of God himself, were nevertheless thought to be somehow closer to the worshipper and believed to have power over limited aspects of life. In this instance however, the tradition of monotheism remained too strong for a complete lapse into polytheism, but the trend is nevertheless discernible, as Clube and Napier were quick to point out.

On the other hand, the suggestion made by these writers that the population at large was directly aware of the sky god or high god (in the form of a bright comet) is open to serious criticism. Without wishing to go too far into the history of religious belief, it seems to me that belief in a deity did not spring from a widespread experience so much as from the teaching of a small number of charismatic, in both the technical theological and popular senses of that term, sages. These people were recognized by the wider population as having experienced conscious states not common to the majority of folk, in which they claimed direct contact with a realm beyond the material. In the most striking instances, they believed that they were receiving knowledge from the Supreme Deity and that it was their role to instruct those around them of this fact. This process is most conspicuously represented by the Hebrew theistic tradition. Although the strongest of the monotheistic cultures, the belief of the people was principally determined by the teaching of a series of prophets, not by the perception of something in the sky. Then, after the age of the prophets passed, God was understood as being increasingly remote. This is reflected in the Hebrew literature written in the period between the Old Testament and the New, where the term "God" is increasingly replaced by the alternative "Heaven"; the latter term being impersonal and suggesting a transcendent realm rather than a divine Person. As the medieval church looked to saints as intermediaries between humanity and this transcendent God, so the Jews of that period looked to angels to fulfill this role. In a culture less steeped in monotheism, this process might have ended in a transition to polytheism as, indeed, had essentially happened at times during earlier Hebrew history. To discuss the nature of the experiences of prophets and mystic would go beyond the subject of this book, although I personally find no difficulty in interpreting them in quite a literal manner. Nevertheless, whether one takes this view or not, experiences of this nature must be taken into serious consideration in any complete assessment of the human condition, as F.C. Happold once remarked in the course of his study of mystical experiences. The prominent role of prophets, mystics and similar figures seems to have played a greater role in the religious history of mankind than any astronomical occurrences.

Of course that does not mean that astronomical events did not, in ancient times, have any impact on the religious life of the population. After all, it remains true that even today ideas are spawned by events that, while not necessarily very rare, are infrequent enough and sufficiently spectacular to arouse wonder and even fear amongst a segment of the populace. No doubt this happened in ancient times as well. Astronomical events of a spectacular kind—whether ones that are familiar to us in our day and age or whether something so rare as to never have been officially catalogued—undoubtedly exercised some influence on human culture down the ages, even if this has not been as fundamental as Clube and Napier believed.

Yet, even if the hypothesis put forward by Clube and Napier concerning a cometary root of religious belief is weak, they may still be correct about a truly giant comet as the progenitor of the Taurid complex and, indeed, about the role that this system has played in catastrophic terrestrial events.

It should be mentioned that a giant comet—in the sense of an unusually massive one—need not necessarily be an abnormally bright object. This might appear counterintuitive, but we need only compare the intrinsic magnitude of Chiron, which Clube and Napier took as a model for their hypothetical giant comet, and Hale-Bopp when at comparable distances from the Sun. The smaller (albeit still unusually large) Hale-Bopp was the brighter and more active of the two, although in all fairness we cannot say how Chiron would behave if it came within one Astronomical Unit of the Sun. Still, even a very massive comet will not develop a bright coma if most of its surface is covered by an insulating crust. It is possible that, because of its comparatively strong gravity—compared with that of a normal-sized comet—something the size of Chiron might build up a crust more readily than a smaller object. As a very large comet approaches the Sun and weak activity begins at several Astronomical Units, dust particles feebly wafted aloft by the escaping gases might be more prone to fall back to the surface than those ejected from a small comet with only very weak gravitational attraction. Furthermore, if there were significant quantities of short-lived radioactive isotopes such as Al 26 present in the material from which the comet formed, a body of Chiron's dimensions may have become sufficiently hot at its center to melt rocky material and form a molten core which would subsequently solidify as the isotope decayed. Although such a hot center would not drive ice away from the outer layers of the comet, an object of this type may have experienced small-scale volcanic eruptions where molten lava could bubble to the surface in certain weak spots and, after solidifying, contribute to a refractory crust.

The question that must be asked however is whether the original progenitor of the Taurid complex—proto-Encke—was really a Chiron-type object. What evidence exists for thinking that it was something larger than an "ordinary" large comet?

Clube and Napier arrive at their giant comet conclusion by adding together the combined mass of all the objects allegedly associated with the Taurid complex. This not only includes the total estimated mass of meteoroids, but the more significant mass of an estimated 100 asteroids together with the contemporary Encke's Comet. If all of these objects were combined together into an approximately spherical body, the diameter of such a body would be around 100 km (62 miles). Approaching the situation from the other direction therefore gives us the dimensions of the supposed parent object that progressively fragmented into the complex that we observe today.

There is little doubt that Encke was a good deal larger, several thousand years ago, than it is today. There is also little reason to dispute the widely-held opinion that a good many sizable lumps of matter—fragments of the ancient Encke that have long since lost their volatiles and now appear essentially asteroidal—lurk within the Taurid complex. But has their number and total mass been overestimated by supporters of the catastrophist role of the Taurids? But is the Taurid complex really as massive as these astronomers figure?

There are reasons for doubt. For instance, it is far from certain that many of the asteroids whose orbits bear some resemblance to that of the Taurid meteors are truly members of the family. Many may be gatecrashers; interlopers with only a coincidental orbital resemblance to the Taurids or to Encke's Comet. Following a study of the similarities between Taurid meteor orbits and the orbits of what could be termed Taurid-like asteroids, J. Klacka concluded that an observational selection effect favors the discovery of asteroids in these types of orbits and that the assumption of a common origin is therefore not warranted. This is supported by the discovery that many of the supposed Taurid asteroids display a reflectance spectrum closer to that of the typical object of the inner asteroid belt than to any comet nucleus, or suspected defunct comet nucleus, thus far observed. The latter objects are very dark, yet a significant number of the asteroids in Taurid-like orbits have far higher albedos, more reminiscent of S-Type asteroids and ordinary chondritic meteorites. For instance, a 2014 study by M. Popescu et al. of the six largest Taurid-like asteroids found that five of these have S-Type spectra. Only one (1996 RG3) seems to be dark and possibly of the C-Type. Therefore, of the six objects studied, the latter appears to be the one most likely to have a cometary origin, although this cannot be taken as proven fact on the grounds of albedo alone.

One of the most convincing candidates for being a true Taurid asteroid is 2004 TG10. The orbit of this object is very close to that of Encke and it seems entirely reasonable to think of the asteroid as a fragment that broke away from Encke sometime in the past. Although dormant or defunct today, 2004 TG10 may well have been an active comet once. In fact, this object, rather than Comet Encke itself, is even suspected of being the *immediate* parent of the northern branch of the Taurid meteor shower, whereas Encke itself seems to be the immediate parent of the southern branch. On the other hand, it may simply be a very large "meteoroid" within that stream!

Of the other possible large Taurid asteroids, the one which for a time appeared to yield the best evidence of a cometary past was Oljato. Clube and Napier backtracked the orbits of Oljato and Encke and found that these came into close proximity some 9500 years ago; a fact which they interpreted as evidence that both objects are fragments of a single body that broke apart at that time. However, that is not the main reason that a cometary past was suspected for this asteroid. More suggestive of this was the discovery of apparent low-level cometary activity hinted at by some curious observations in the 1980s made by the Pioneer Venus Orbiter. This spacecraft observed three passages of the asteroid between Venus and the Sun and on each occasion it recorded a marked increase in peaks of a type of unusual magnetic disturbance known as Interplanetary Field Enhancements or IFEs. These occurred both ahead of and behind the asteroid and were interpreted by some as evidence that Oljato was really a very weakly active comet. Visually, its appearance was asteroidal, but the presence of the IFEs seemed explicable in terms of a low level of dust emission from the body into what might best be called a subvisual dust tail. Presumably, the object would have been far more active in the more or less distant past and may once have been a rather bright comet. Indeed, Clube and Napier's orbital computations indicated that it would have been in a position to make close passes of Earth during an interval of several centuries around 3000–3500 BC and may then have been a conspicuous object during favorable apparitions.

Nevertheless, the suggestion that Oljato was a comet was not without problems. Like some other supposed Taurid asteroids, it did not have a typical cometary color. Its reflectance spectrum is somewhat odd, but it is known to be around ten times more reflective than any of the comets thus far studied. That alone makes it more compatible with an interloper from the inner asteroid belt and indicates that the calculated convergence of its orbit with that of Encke 9500 years ago may be nothing more than pure coincidence.

It may, however, be possible to rescue Oljato's hypothesized Encke/Taurid connection on the grounds of a suggestion made earlier that giant comets may possess rocky cores. If that is correct it is possible that objects not unlike S-Type asteroids—fragments of just such a rocky core—might be found embedded within certain cometary meteor streams. If the Taurid complex really is a meteor stream originating from the disruption of a giant comet, it could be suggested that Oljato and the other reflective asteroids in the Taurid complex originated from the breakup of its core. On the other hand, if a core of this nature did exist in the Taurid comet, it may have been quite small and, as even its very existence is far from established this suggestion should be considered only if all else fails.

Nevertheless, even more puzzling observations of Oljato were to follow. When Venus Express arrived at the planet in 2012, it failed to observe any IFEs associated with passages of the asteroid. Not only that, the rate of these disturbances in the regions immediately behind and ahead of Oljato were actually lower than average! What was going on here? Had the comet given up the proverbial ghost sometime between 1980 and 2012? After 9500 years of presumed activity at some level of intensity, it would seem a little strange that we just happen to have caught its last gasp.

A more convincing explanation, and one that did not involve cometary activity, was given by Dr. C. Russell. According to Russell,

> At one point in time Oljato shed boulders—mostly a few tens of meters in diameter—into its orbit and they formed a debris trail in front and behind Oljato. These impactors then hit other targets as they passed between Venus and the Sun. The large amount of fine dust released by these collisions was picked up by the solar wind, producing the IFEs observed by Pioneer and was accelerated out of the Solar System.
>
> The reduced rate of IFEs observed during the Venus Express epoch suggests that the collisions with Oljato's co-orbiting material have reduced the general debris in the region as well as the co-orbiting material shed by Oljato.
>
> The IFEs observed by Pioneer suggest that more than 3 tonnes of dust was being lost from the region each day. Effects associated with solar heating and gravitational perturbations have gradually nudged larger chunks of debris from Oljato's orbit. From once being unusually crowded, the region has become unusually clear and free of IFEs.

Presumably, the event causing Oljato to shed boulders happened quite recently and may have been the result of a collision with a large meteorite or even thermal stress following many passages within the orbit of Venus. The real point, however, is that this does not involve Oljato ever having been a comet in the true sense of that word, even if it may possibly have presented a temporary "cometary" appearance at the time of the disruption, as other asteroids have been observed to do in recent years. Nothing here requires an association with either Encke in particular or with the Taurid complex in general.

Another observation which a number of astronomers, Clube and Napier amongst them, have cited as evidence for swarms of relatively large bodies within the Taurid complex—bodies large enough to pose as "thunderbolts" if not necessarily to cause widespread destruction—is the anomalous peak in the number of large meteorites striking the Moon between June 22 and 26 in the year 1975. For these 4 days, a cannonade of bodies as massive as 1 ton struck the surface of the Moon. These objects approached our satellite from the sunward side and apparently coincided with activity in the Earth's ionosphere which may have been indicative of enhanced meteoric activity. June is the month of the daytime Beta Taurid meteor shower which is known to be a branch of the broader Taurid complex in which the meteoroids strike Earth as they travel outward from the region of the Sun. The shower is most active late in the month, although the rates are always rather low. It is this coincidence of date and direction (approaching the Earth/Moon system from a more or less solar direction) that implicates the Beta Taurids in the minds of Clube, Napier and quite a large number of other astronomers.

Nevertheless, the case is certainly not open and shut. If one wished to play Devil's advocate on this matter, it is easy enough to draw attention to another June shower with a radiant not too far from that of the Beta Taurids, albeit having an entirely dissimilar orbit. The meteors of this shower also approach from a direction not too far from the Sun and arrive on Earth during the daylight hours. Moreover, although the peak of this stream comes just before the middle of the month, it is more intense than the Beta Taurids and activity extends past the end of June and into early July. The members of this shower radiate from the constellation of Aries; the celestial neighbor of Taurus.

The June Arietids form an interesting stream. Like the Beta Taurids, they too are part of an extensive complex of cometary debris. Indeed, it has been known for several decades that the shower is associated with the Delta Aquarids of July. Moreover, backward computations of the orbits of the well known January Quadrantid display rather surprisingly revealed that this shower, although now pursuing a path very different from that of the Delta Aquarids, actually had an almost identical orbit to this shower about one thousand years ago!

For a long time no parent object could be found for the Arietids or Delta Aquarids. The meteoroids pass unusually close to the Sun and for a long time no corresponding comet was known. Sekanina at one time suggested that the asteroid Icarus might be the parent, not that its orbit bore a very close resemblance to that of the meteor stream, but simply because it was then the only body known whose orbit was even remotely similar.

Then, during the opening years of this century, the mystery started to clear. Examination of data from the space-based SOHO coronagraphs collected in previous years uncovered the existence of several tiny comets having perihelia somewhat larger than the true sungrazers albeit still unusually small. Astronomer Brian Marsden found that the observed arc of several of these could fit essentially the same orbit, although the times of their appearance ruled out different returns of the same object. Clearly, another comet group had been discovered and this has justifiably been given the name of the Marsden group. One of the Marsden comets, recorded in May 1999 but not actually discovered until a couple of years later in archived SOHO data, was somewhat brighter than the others and according to Marsden's calculated orbital elements must have made a very close approach to Earth around June 12, approximately 1 month after its perihelion passage. Of course, nobody knew anything about this at the time and the very faint and fast-moving object passed us by unnoticed. It was then that the present writer noticed an apparent similarity between the orbit of this comet and that of the Arietids and notified Dan Green at the Center for Astronomical Telegrams of a possible association. Further data has strengthened this link—always an encouraging sign! Subsequently, yet another grouping of SOHO comets was noted by Reiner Kracht and this one was also linked by Marsden with the Marsden group and the Arietid/Delta Aquarid/Quantrantid meteor complex (yet another SOHO comet group was discovered by Maik Meyer, but this one is not part of the complex and is, as they say, another story).

The SOHO comet of May 1999 returned in 2004, this time accompanied by a small companion comet travelling in essentially the same orbit but slightly ahead of the main object in terms of its time of perihelion passage. A study by Sekanina determined that the smaller object had broken away from the main body early in 1999, several months prior to the May perihelion. The two objects were too close together for SOHO to discern them as separate objects when they moved through the coronagraph field that year. They would have been enveloped within a single coma at that time, but slowly separated into two comets as they withdrew from the Sun. Using the upgraded elliptical orbit, the two comets passed just 0.0087 AU from Earth and 0.0091 AU from the Moon on June 12, 1999. The principal comet was observed again, both in SOHO and STEREO images, in 2010 but the secondary was not recovered and may have continued to fragment into ever smaller pieces.

It is now widely accepted that both the Marsden and Kracht groups of SOHO-detected comets, the Quadrantid, June Arietid and Delta Aquarid meteor showers, one asteroid (more likely a defunct or dormant comet nucleus), possibly one historical comet (C/1490 Y1) and another tiny comet discovered by Kracht in SOHO images and apparently following an orbit very similar to that of the Delta Aquarid meteor stream, are all part of a vast debris complex associated with the peculiar short-period comet 96P/Machholz, formerly known as Comet Maccholz 1. It is interesting to note the addition of two further members of the complex in 2012. Comet Machholz returned in July of that year and had apparently shed a fragment at some time since its previous perihelion passage. This fragment subsequently split into two pieces which returned as a pair of tiny comets accompanying the main one through the 2012 perihelion. The small objects were not seen from the ground, but were discovered in SOHO images, published on the Worldwide Web, by Prafull Sharma (at the time a secondary school student only 16 years of age) and independently by Liang Liu. Whether Machholz is breaking up into another comet group within the broader complex, whether this was just an isolated event and whether either of these small Machholz group comets will be recovered during future perihelion passages remains to be seen.

For the present purpose however, it is interesting to keep in mind that the Arietid stream is known to be associated with progressively fragmenting objects several tens of meters in diameter, so the conjecture that fragments of one of these struck the Moon in 1975 is not at all farfetched. We can, I think, make just as strong a case that the Arietids were responsible for that event as we can make for the Beta Taurids.

On the other hand, it may be that neither the Machholz complex not the Taurid complex was responsible. A study by J. Oberst and Y. Nakamura of the 1975 event—together with a similar event that occurred in January 1977 (for which, incidentally, the Taurids could hardly be blamed)—concluded that both meteorite swarms may have included metallic objects. If that is correct, it would appear that asteroidal fragments were responsible and that neither the Arietids not the Beta Taurids can be implicated. The formation of iron meteorites requires conditions typically found in relatively large differentiated bodies and it is very difficult to believe that they could originate in comets; not even in giant members of the species.

Another event which Clube and Napier associate with the Beta Taurids is the famous Tunguska bolide of June 1908. Much has been written about this event and all manner of weird and wonderful suggestions as to its true nature have been put forward over the years (the suggestion that it was an antimatter meteorite, discussed earlier, being just one of these—and not even the weirdest!). Several possible orbits have been calculated from the rather meager information concerning the path of the meteor and some of these are not too dissimilar from those of the Beta Taurids. Nevertheless, as noted earlier, Sekanina's study of the event concluded that the orbit of the body was quite typical of that of a small Apollo asteroid originating within the inner asteroid belt. Such a body would presumably be an S-Type asteroid; a class of object thought to be of similar composition to the most common type of meteorite, the ordinary stony (chondritic) variety. In apparent confirmation of this, Sekanina also found that the degree of atmospheric penetration of the body suggested a tensile strength similar to that of ordinary chondrite meteorites. Moreover, meteoritic dust, of a type consistent with the composition of ordinary chondrites, has been found in the soil in the region of the event. These results make any association with the Taurid complex seem extremely unlikely, despite the coincidence of dates.

The final event occurring within historic times which has been widely seen as evidence for large and potentially destructive bodies embedded within the Taurid complex is the report by a group of Canterbury monks of a remarkable occurrence early in the evening of June 18 in the year 1178. Imagine the scene. A warm summer's evening and a clear western sky in which hung a bright crescent Moon. Then, without warning, the Moon appeared to experience a series of convulsions unlike anything reported either before or since. In the words of Gervase of Canterbury;

In this year, on the Sunday before the Feast of St. John the Baptist, after sunset when the Moon had first become visible a marvelous phenomenon was witnessed by some five or more men who were sitting facing the Moon. Now there was a bright new Moon and as usual in that phase its horns were tilted toward the east and suddenly the upper horn split in two. From the midpoint of this division a flaming torch sprang up, spewing out, over a considerable distance, fire, hot coals, and sparks. Meanwhile the body of the Moon which was below writhed as it were, in anxiety, and to put it in the words of those who reported it to me and saw it with their own eyes, the Moon throbbed like a wounded snake. Afterwards, it returned to its proper state. This phenomenon was repeated a dozen times or more, the flame assuming various twisting shapes at random and then returning to normal. Then after these transformations the Moon from horn to horn, that is along its whole length, took on a blackish appearance. The present writer was given this report by men who saw it with their own eyes, and are prepared to stake their honor on an oath that they have made no addition or falsification in the above narrative.

After languishing forgotten for centuries, this record was brought to light again by Paul Hartung in an article in the journal *Meteoritics*, published in 1976. Hartung argued that what the Canterbury monks actually saw was a small asteroid striking the Moon. This interpretation was furthermore supported by two pieces of circumstantial evidence. First, the reported location on the Moon of the flaming torch appeared to be consistent with that of the crater Giordano Bruno, which appears to be the youngest of the Moon's major craters and, secondly, laser-ranging experiments carried out during the 1970s revealed the Moon to be oscillating in such a way as to suggest that it had been struck by a large projectile several hundreds of years ago and was still reverberating from this impact. In view of this, the impactor hypothesis did not appear to be all that farfetched. Moreover the date (18 June) also looked interesting in view of Beta Taurid activity; a fact quickly noted by Professor D.I. Steel and taken up by Clube and Napier. Recently, D.W. Chapman has drawn attention to the global cooling that took place in the late twelfth century, arguing that this may have resulted from large quantities of Moon dust, ejected by the impact, entering the upper atmosphere of Earth (Fig. 4.8).

Nevertheless, the lunar-impact explanation of the 1178 event, though presented almost as a self-evident truth in several astronomy books and articles, has some serious problems. For one thing, it is surely strange that such a spectacular and (quite frankly) frightening event should have been witnessed and/or chronicled by only one small group of people at one location. Didn't anybody else see it? Furthermore, the size of the torch and the darkening effect that spread over the Moon, if truly at lunar distances, implies

FIG. 4.8 Giordano Bruno lunar crater (*Courtesy*: NASA)

unrealistically high velocities of ejected particles. And then what are we to make of the contortions of the flame and the dozen or so repeat performances of the phenomenon? Or the throbbing of the Moon "like a wounded snake"? Quite frankly, these portions of the account do not read like the descriptions of an impact on the Moon, yet if we are going to take the account seriously, we must assign an equal weight to all of its aspects and avoid any temptation to cherry pick only the parts that agree with our interpretation.

If the account of the event includes certain matters that do not nicely fit with the impact interpretation, it can equally be said to exclude other aspects which we might expect if that interpretation was the correct one. For instance, an impact of the size being postulated here must have kicked up a lot of fine dust and one would think that this dust plume would have been visible following the event. More seriously however is the question of what is supposed to have become of the matter blasted out of the crater. According to calculations by Paul Withers of the Lunar and

Planetary Laboratory, large amounts of this material would have struck the atmosphere of Earth in what should have been the father and mother of all meteor storms; something like the 1966 Leonids but lasting for at least a week! Such an event would have struck terror into the hearts of people of that time. I dare say that it would strike terror into the hearts of many folk in our time as well. Yet, nowhere is such an event mentioned. The only reasonable explanation for this omission is that no such meteor storm eventuated. Withers also casts doubt on the idea that the 1970s laser-ranging results supported a relatively recent large-scale impact. By reanalyzing these data, he concluded that the slight oscillation detected arises from fluid motions deep within the Moon's interior and not from the ringing following a powerful impact.

The very young age of the Giordano Bruno crater has also been called into serious question. This was always a contentious issue with certain lunar experts. For instance, the well-known British lunar observer, Sir Patrick Moore, insisted from the start that the crater, although very young by the standards of lunar formations, was formed a lot earlier than 1178. Recent observations have apparently proved the skeptics right. From a count of small craters within the ejecta blanket of Giordano Bruno, as observed with Japan's SELENE lunar orbiting spacecraft, T. Morotu and colleagues estimate that the real age of this formation is somewhere between one and ten million years. Other estimates by, for example Y. Shkuratov and V. Koydash, also placed the crater's age at one million years or more. Clearly, the Giordano Bruno crater was not formed in 1178 (Fig. 4.9).

But if the Canterbury monks did not witness an asteroid striking the Moon, what did they see?

The most likely explanation is that they witnessed a bright exploding meteor, within Earth's atmosphere, that just happened to lie in the exact line of sight with the Moon. It might even have been a point meteor, that is to say, they might have been located at the exact spot on Earth's surface where the meteor would appear to come head-on from its radiant position. Had the moon not been present, they would have seen the meteor begin as a point of light and grow into a spectacular exploding fireball whilst not appearing to move in either altitude or azimuth. If the object exploded several times during its trajectory

Fɪɢ. **4.9** Giordano Bruno crater imaged from Apollo 14 (*Courtesy*: NASA, Lunar and Planetary Institute, Apollo Image Atlas Hasselblad Image Catalog AS14-75-10306)

through the atmosphere, the multiple appearances of the torch can be explained and the distortion and eventual darkening of the Moon's image was almost certainly due to the eyewitnesses' view down the entire length of the train of dust and expanding gases left behind as the meteoroid disintegrated. Because one needed to be at the exactly right spot at the right time to witness this effect, anybody outside Canterbury would have simple seen a bright meteor.

This might seem like a grand anti-climax. The hypothesis that the Taurid meteor stream was once responsible for the rise of beliefs and the fall of civilizations seems to rest on very shaky ground. So does the fear that it all might happen again; that a hypothesized stream of large bodies forming a sort of core within the Taurid complex will one day cross Earth's path again and bring all manner of disaster to our planet. Certainly, there are large bodies within the complex, but the idea that this system of debris hangs over us like the sword of Damocles is an overstatement. The Taurid complex appears to be well down on our list of things to worry about.

The Celestial Swastika?

If there is a symbol which strikes dread into the heart of most people living since the 1930s, it is surely the swastika. Its very mention evokes thoughts of genocide, racial hatred, fanatical Right-wing political ideas and a raving megalomaniac who brought the world into the worst conflict that the human race has yet endured. Words like "evil", "diabolical" and even satanic easily rise in the consciousness at the very sight of the crooked cross. Yet, this revulsion is a recent thing. Before the symbol was hijacked by Hitler's party of national socialists (fascists in the estimation of everybody else) it enjoyed a very long history as one of the most enduring of all symbols. In India, it was used by the Jain, a religious movement whose reverence for all life was about as far removed from Nazi philosophy as it is possible to be.

The word swastika comes from the Sanskrit svastika and has the meaning of good fortune or well being. But the symbol itself preceded the term. Swastikas appear in the art of Eurasia, as far back as high Neolithic times dating between 4500 and 3500 BC. Over the centuries and millennia the symbol has appeared in Scandinavia, India, China, amongst the Maya of Central America and the Navajo of North America and in the form of the crux gammata (a cross consisting of the joining of four Greek gammas) in early Christian/Byzantine art.

Although it has in most instances represented wishes for good fortune, it has also acquired a number of different meanings in certain cultural traditions. Thus, the Jain of India, among whom it is a major symbol, see the swastika not only as a symbol of good fortune but also, because of its possession of four arms, as a reminder of what they believe are the four possible post-mortem destinations of Man; rebirth into the animal or plant realm, descent into Hell, reincarnation back into the human realm or ascent into the spirit world. Buddhists see it as representing the footprints of the Buddha himself.

For the most part, the arms of the swastika are shown as pointing in a right-handed manner, such that, if the symbol is rotated, it would naturally turn in a clockwise direction. At times, however, it is represented as pointing in a left-handed or counter-clockwise direction and on at least some occasions this appears to deliberately imply a more sinister (in both the literal and metaphoric sense) message. Thus, in ancient Scandinavia, the left-handed swastika stood for the hammer of the thunder-god Thor and among the Hindus and Jain the left-handed swastika is a symbol of night and darkness as the right-handed version stands for daytime and light. Even more sinister, the Hindus associate the left-handed swastika with Kali, the goddess of destruction. Some indication of the activity of this goddess is given by the fact that her devotees, known as Thuggees or Thugs, have given the English language the term "thug" for people whose behavior displays similar traits, albeit without the theological implications. This form of swastika is also associated with the practice of magic and dark arts. By the way, despite attempts to show that the Nazis deliberately used the swastika as a satanic symbol, theirs was actually the right-handed one.

The widespread and ancient use of the swastika symbol is one of those mysteries that have inspired all manner of hypotheses over the years. Many have seen an astronomical association and, indeed, the prevalence of the right-handed version of the symbol and its association with light and daytime is widely thought to be representative of the right-handed motion of the Sun across the sky. Perhaps it in some way represents the Sun, although the four bent arms have no obvious solar association. Others have suggested patterns of neural activity inducing subjective swastika-like patterns of light. Yet another suggestion is that it represents the claw-marks of birds. Perhaps all of these suggestions possess an element of truth, but something happened late in 1973 which really stirred up speculation, and not a little controversy.

That year, in the village of Mawangdui in the Hunan Province of China, a number of books written on silk scrolls were discovered in a tomb of a man who had been the chief minister's son back in the year which our calendar marks as 168 BC. From other evidence found amongst the funerary goods, it seems that this gentleman died on April 4 of that year, as our system of time reckoning would declare. One of the documents discovered bore a title which, as translated into modern English, read *Prognostications Related to Astronomical and Meteorological Phenomena*. Subjects covered were halos, clouds, rainbows, mirages and comets and the supposed way in which such phenomena could be used as portents.

One page of this book has become known as the Comet Atlas and contains drawings of 29 different forms that these objects may take, together with brief comments concerning the supposed astrological significance of each of these. Most of the forms represented are relatively easily identified with those assumed at various times by different comets and the majority of these are given only a short commentary in the text. There is, however, one exception. One of the comet forms depicted in the Atlas is a swastika. The proportions may not be quite the same as the usual representation of this symbol (the arms—especially the vertical ones—are a little longer and their respective lengths not so regular as most drawings of the figure) but the image is nevertheless unmistakably a swastika. It also turns to the right, although according to the astrological commentary associated with this image, only one of its supposed effects is actually beneficent. Translated into English, the commentary informs us that when this form of comet is "seen in Spring [it] means good harvest, seen in summer means drought, in Autumn means flood, in winter means small battles."

What is to be made of this? Why should a comet be depicted as a swastika? Or should the question be turned around and should we ask if the swastika is really the representation of a particular form of comet? Before even attempting any answers, let's look at a few salient points in the hope that some clue might be supplied.

First of all, note that the commentary concerning this comet-form is the longest and most comprehensive in the Atlas and it is the only one that draws distinctions between different seasons of the year in which comets having this form appear. From this fact alone, it would seem that this form of comet had been observed many times and at all seasons prior to the date of the Atlas' composition, which must have been prior to 168 BC. This is something to which we shall return later.

Secondly, it is interesting that the name given in the Atlas to this form of comet—Di Xing—translates as the long-tailed pheasant star. The word Di refers to a kind of Tartar pheasant noted for having a long tail. This immediately raises the question as to whether the long-tailed in the name of this comet-form implies that the star itself had a long tail or whether it in some other way related to the bird known for its long tail. From the depiction of this form of comet, it does not seem that a long tail characterized its appearance. Of all the comet-forms represented in the Atlas, the Di Xing is the only one represented as being devoid of tail. Or, maybe, it is not necessarily devoid of tail per se, but the presence or absence of a tail is not the feature to which the Atlas' compiler wished to draw the reader's attention. The characteristic of this form of comet was not the tail, but the peculiar appearance of the head. The question of the long-tailed pheasant association will also be taken up again later.

A third point to remember is that, although the swastika became a recognized symbol in China during the Common Era, scholars think that it was initially introduced to that country from India as the Buddhist faith spread beyond the bounds of the land of its birth. (A swastika-like sign found in pottery from Chinese Neolithic culture dating back to 2400–2000 BC appears unconnected with later usage of the symbol.) However, despite legendary tales of Buddhism in ancient China, the consensus amongst scholars is that this religious and philosophical system did not reach that country until the first century AD. If the swastika, as a religious or oracular symbol, was not known in China until brought there by Buddhist missionaries and travelers and if this contact with Buddhist beliefs and symbolism did not take place until the first century of our era, it follows that the symbol would have had no metaphysical significance to the compilers of the Comet Atlas in the years before 168 BC. This adds to our confidence that the form depicted really did represent, more or less faithfully, something truly observed in the sky and was not a pre-existing symbol psychologically projected onto something that only vaguely resembled that form.

This, however, throws up another puzzle. As was said earlier, the relative length of the commentary concerning Di Xing and the mention of its changing astrological significance according to the time of year when it appears strongly implies that many instances of this form of comet had been noted over the years and that these apparitions occurred at all seasons. Without giving credence to astrology, it remains true that as practiced by the ancient Chinese this was an empirical pursuit. If the ancient astrologers came to associate a certain type of astronomical event with something happening on Earth it was because they believed that they had evidence that the one caused the other. Sometimes this supposed evidence may have related to the form of an astronomical object (a broom star comet, for instance, was thought to sweep evil from the heavens and down onto the Earth) but the most convincing evidence was surely the correlation between a specific type of astronomical event and an equally specific type of terrestrial one. If there were several instances of the Di Xing appearing in winter, for example, and most or all of these were followed by small battles, a correlation appears to have been found. But correlations require more than just a single example, and because the Di Xing is correlated with different events at each season of the year, it would seem that many examples of this comet were indeed observed and that these sightings were spread throughout the year. The mystery raised by this however, is that relatively comprehensive records of comets observed in China extend from the beginning of the Common Era (with some scattered entries from earlier times) until just over a century ago, and yet the Di Xing form is nowhere mentioned. This is even more surprising considering the fact that this is also the period when Buddhism spread throughout China, carrying the swastika symbol with it.

Carl Sagan and Anne Druyan appear to have been the first to draw attention to the swastika-like appearance of the Di Xing and, in their 1985 book *Comet*, offered a possible solution of the mystery. They drew attention to the more-or-less swastika-like spiral forms sometimes assumed by jets emerging from active spots on the rotating nucleus of a comet, especially if observed along the comet's axis of rotation. In particular, they pointed to the swirling patters within the inner coma of Comet Bennett in 1970.

Nevertheless, there was nothing obviously reminiscent of a swastika about Comet Bennett as observed with the naked eye. Relatively large telescopes with high magnification and photography revealed patters somewhat similar to the swastika, but that is hardly relevant to an astronomer/astrologer in ancient China observing with his eyes alone. If these patterns are to be visible at all without optical aid, a bright and active comet must pass very close to Earth, and it was exactly this scenario that Sagan and Druyan proposed. According to these authors, at some time in the rather distant past, a large comet came extremely close to Earth and for a time blazed in the sky more brightly than any object except the Sun. This hypothetical comet was rather like Bennett in that

it possessed several well-defined jets and its axis of rotation fortuitously happened to lie along the line of sight of Earth's inhabitants, such that the inner coma region spectacularly displayed these
jets in the familiar swastika pattern. So bright was this comet that
the silvery swastika even blazed in the daytime sky, witnessed by
a large part of the world's population as it then existed and, from
these early witnesses, passed into legend and symbol.

This is all very colorful, but does it do justice to the facts?
Even overlooking, on the grounds that very unlikely events do
sometimes happen, the low probability of a suitable comet having the right axis of rotation coming so close to Earth under just
the right circumstances for such a spectacle to be seen, the Sagan-
Druyan scenario does not obviously account for the "seasonal
variations" in the astrological significance of the Chinese account
of Di Xing. Whether or not a single close encounter by the hypothetical swastika comet would have been capable of inspiring the
swastika symbol across much of the human race, it could hardly
explain the supposed differing significance of its appearance during
each of the four seasons unless we imagine repeated close encounters at all times of the year. Given the type of encounter that Sagan
and Druyan envisage, that is surely stretching credibility.

Another problem is raised by the rapid movement of a comet
passing so close to Earth. What was actually seen would depend
to a large degree on where one was situated on the surface of the
planet as the comet rushed by. We might speculate, not altogether
flippantly, that if a large cloud system was present at the time, the
swastika symbol might never have existed.

If we forget, for the moment, about the wider occurrence
of the swastika symbol and simply think about the Di Xing, the
idea of a single comet of short period does seem more plausible.
The Chinese were careful observers and did not require the blaz-
ing daylight object suggested by Sagan and Druyan to draw their
attention. Neither did they see this form as a symbol. They simply
recorded what they saw and attempted to work out its significance.
A comet of short period moving in an orbit with both ascending
and descending nodes close to Earth's orbit could make close passes
of our planet both on its way to perihelion and also receding from
it. Halley's Comet, for example, has made close approaches late in
the year prior to perihelion on some returns and close approaches

early in the year following perihelion on others. It is theoretically possible for a comet to be so placed that close approaches in (say) winter could occur during certain returns and similar approaches in summer could happen on others. If perihelion lay just within Earth's orbit, it would be possible for such a comet to remain in relative proximity to Earth for quite an extended time. It might just be possible for it to be seen at rather close range at sometime during each of the four seasons.

It is not impossible that a periodic comet having this sort of orbit did exist in ancient times. Apart from the very fortuitous nature of its orbit, we need not imagine that there was anything very unusual about it. It need not have been unusually bright, for instance. At least, not if we only need to account for its presence in the Chinese Atlas. All that is required is the presence of a series of strong jets issuing from the nucleus and an axial orientation that gave the early astronomers a pole-first view of this object. This might be an improbable array of circumstances though not, I think, an impossible one.

On the other hand, I am inclined to think that another scenario may be more likely. Once again, this involves a comet of short period, but in this instance we suggest one moving in an orbit of very low eccentricity with perihelion beyond the orbit of Earth. An object of this nature would, in theory, be visible all around its orbit and most readily observable when at opposition, although we are not suggesting something intrinsically bright enough to be constantly visible with the naked eye. Yet, suppose that this comet was given to super outbursts of brightness, like that experienced by Comet Holmes in 2007 when it suddenly changed from being a faint telescopic object to become a conspicuous naked-eye spectacle. Clear structure was also visible within the coma of Holmes, although this did not assume a swastika form. But suppose this hypothetical ancient comet did assume this form during outbursts. That is surely something that the ancient Chinese astrologers/astronomers would have noted. Moreover, because outbursts of a comet in an almost circular orbit can occur at any point within that orbit (as demonstrated in our own day by the periodic comet Schwassmann-Wachmann 1) they could occur at any season of Earth's year, neatly accounting for the noted seasonal variation in its astrological significance.

If the Di Xing was an outbursting comet, most likely the out-bursts would have been concentrated in a single active region. This does not obviously explain the four arms of the swastika form; however it is possible that such a pattern could emerge if an active region located on a very irregular-shaped nucleus passed in and out of sunlight, giving bursts of activity, as the nucleus rotated. These outbursts, which may have been very frequent as implied by the relatively long commentary in the Atlas, undoubtedly took their toll on the comet and we might suppose that it broke up and disap-peared before the regular records of comets and other astronomical phenomena that are now in our possession were made. Records of specific apparitions of Di Xing may indeed have been made, but were lost when Emperor Ch'in instigated the burning of the books (together with any of their authors who were still around) in 213 BC. It is a good thing that the traditions concerning comets and other omens dating back before this unfortunate event were pre-served in this rare book and then safely tucked away in a tomb for posterity to uncover.

Why was this form of comet associated with the long-tailed pheasant? Pure speculation of course, but recall our earlier men-tion that one explanation of the swastika symbol concerns its apparent similarity with the footprints of birds? Maybe it simply reminded the early Chinese observers of the claw marks left by the long-tailed pheasant. The naming of this form of comet would in that case be no more or less significant than the term woolpack often given to large cumulus clouds.

If this line of speculation is correct, the presence of the Di Xing in the Atlas might be explained, but the mystery of the swas-tika as a widespread symbol remains. Its association by Sagan and others with comets largely rests on the Di Xing, yet it seems that this was understood by the Chinese as nothing more than one of several forms that comets could assume (albeit an important one in very early times, it would seem). They do not appear to have taken it up as a symbol or given it any greater significance than any other astronomical transient. The swastika, as a symbol, did not appear in China until the spread of Buddhism to that land, hun-dreds of years after the Di Xing was drawn in the Atlas and, if our speculations about the nature of this object are in any way close to the truth, long after the Di Xing itself had disappeared from the skies. As the Atlas had already lain forgotten in a tomb for hundreds of years and any other writing that might have described the Di Xing had long ago gone up in flames courtesy of Emperor Ch'in, it is probable that this strange symbol introduced by the Buddhists was not seen as having any association with any type of comet. Conversely, those cultures which have incorporated the symbol from the earliest of times do not obviously associate it with a comet. Claims to the contrary by Sagan, Druyan and some other recent writers, while certainly interesting and thought-provoking, require a lot more evidence before they become truly convincing.

So does the swastika symbol really have an astronomical association? This question will doubtless continue being the subject of debate and, as in the past so in the future, this debate will no doubt range from the sensible to the bizarre. All we can say is that there does appear to be some association between the swastika and the Sun, although it is not immediately obvious why a Sun symbol should be represented as having bent arms. Maybe the bent arms are meant to represent motion, symbolically depicting the Sun in the form of a wheel rolling across the sky in the direction depicted by the arms; the wheel of life so to speak. However, whilst the solar connection likely forms part of the answer, if the mystery is ever solved to the satisfaction of all, my guess is that it will likely also include a psychological dimension. Whilst I do not go all the way with C. Jung's theory of psychological symbolism, he was surely correct in pointing to our use of the number four as a symbol for completeness or totality. We speak about the four corners of the Earth whilst realizing that this makes no sense if taken literally. The four points of the compass similarly depict the entirety of our surroundings. And this says nothing about phrases such as a "square meal" (very few of which are literally square), "on the square", "a square deal" and so on. It seems but a small step to depict the Sun, on which the totality of life depends, as having four arms and an equally small step to bend the ends of these in a way that depicts its motion across the sky. But to proceed any further into the realm of psychological symbolism would require another book.

5 Weird Theories of a Weird Universe

The Cosmic Redshift and Electrostatic Repulsion

An atom that has its full quota of electrons, that is to say, a number equal to that of its protons, is normally considered to be neutral. Any imbalance in these numbers converts the atom into an ion; an anion if it has gained extra electrons and therefore possesses an overall negative charge or a cation (or kation) if it has lost one or more electrons and gained a positive charge. But if the balance of protons and electrons is just right, common sense appears to declare that the atom is neutral. Yet, this "common sense" conclusion rests upon the assumption that the positive charge of the proton is exactly matched by the negative charge of the electron. This is a good assumption, but it is, nevertheless, still an assumption. There is no intuitive a priori reason for thinking that the charges must be exactly equal.

Indeed, if this assumption turns out to be correct, it actually uncovers another of nature's mysteries. The problem is that protons and electrons are two different types of particle. The proton is a composite fermion, consisting of quarks, and it is the particular type and combination of these constituent quarks that give it its electric charge. On the other hand, the electron is an example of a type of particle known as a lepton. These particles, according to the standard model of particle physics, are quite distinct from quarks and particles consisting of quarks, even though both are types of fermion. Quarks feel the strong nuclear force whereas leptons do not. In one sense, it is almost as if they exist in two

D. Seargent, *Weird Astronomical Theories of the Solar System and Beyond*, Astronomers' Universe, DOI 10.1007/978-3-319-25295-7_5

different universes, each invisible to the other! Nevertheless, for these two particles from such dissimilar stables to carry even approximately the same charge must imply that they have some deep connection at high energies and it is in the understanding of this connection that the solution of the apparently simple observational fact that atoms are at least almost neutral (close enough for the neutrality assumption to be made almost without giving the matter any thought) is made.

As we shall see in a little while, experiments conducted over recent decades have narrowed down any possible difference in the charge carried by these two types of particle to an almost infinitesimal degree. In fact, all experiments thus far conducted have yielded results that are consistent with the charges being exactly equal. The equality assumption seems to have been correct all along.

Nevertheless, before these relatively recent results, the hypothesis that a slight difference in charge between protons and electrons formed the basis of an intriguing and ingenious hypothesis within the wider framework of the Steady State cosmology. Nowadays, the Steady State has faded into history, but back in the 1950s and first half of the 1960s (that is to say, prior to the Penzias-Wilson discovery of the microwave background in 1965) it was held in higher esteem than the rival Big Bang theory by most members of the cosmology community. The suggestion that a slight difference might exist between the charges of protons and electrons was published in 1959 in a paper by two of the Steady State theory's strongest supporters, mathematicians and theoretical astronomers R.A. Lyttleton and H. Bondi.

The latter author was one of the trio of cosmologists (together with Tom Gold and Fred Hoyle) who first put forward the Steady State theory. Unlike the evolutionary ("Big Bang") model of G. Lemaitre and G. Gamow, the Steady State proposed that there was no such thing as evolution on a cosmic scale. In other words, in terms of its appearance on a large scale, the universe had always and will always appear essentially the same as it does today. Of course, individual stars and galaxies form, age, and die, but the universe itself always looks the same. In the Steady State scenario, the universe is infinite in extent, eternal in time and Euclidean in its geometry. Yet, being steady state does not mean static. Unlike some of the more radical cosmologies put forward during the first half of last century, the Bondi/Gold/Hoyle model accepted the red-shift in the light of galaxies as evidence of recession. Other theories had been postulated suggesting that this phenomenon was caused by light losing energy as it traversed vast reaches of space over great periods of time or introduced some other factor that was thought to cause light from distant and ancient sources to appear red-shifted. These ideas were ingenious, but the proposers of the Steady State theory accepted the most straightforward explanation that most external galaxies are red-shifted because they are racing away from one another with velocities that are proportional to their mutual distance. As a necessary corollary of this, the theory implied that matter must be being continually created. Even if matter per se is eternal, the expansion of the observable universe means that at a finite distance from any observer within that universe, the velocities of recession of galaxies will reach that of light itself. For all practical purposes, galaxies that are receding at velocities greater than that of light have ceased to have any physical significance to an observer within the radius of the observable universe. If no matter is flowing in from beyond the horizon (a situation precluded by the universal expansion) new particles must be being continually created by some process throughout the volume of the observable universe. Otherwise, the observable universe would have already emptied itself.

But why is the universe expanding? What is driving its galaxies away from each other? The rival Big Bang theory has no problem with this. The universe is simply swelling outward from the initial explosion that initiated the whole story. But if there had never been such a cosmic explosion, the recession of galaxies lacks an obvious cause.

It was at this point that the hypothesis of Lyttleton and Bondi came forth. These scientists found that if the electric charge of the proton and electron differed by as little as two parts in 10^{18}, assuming an equal number of protons and electrons in the observable universe, this minute inequality of electric charge would result in electrostatic repulsion overcoming gravitational attraction on the large, cosmic, scale. The expansion of the universe is, in the final analysis, driven by the same phenomenon of electrostatic repulsion that opens the leaves of a gold-leaf electroscope.

Project 5.1: A Simple Demonstration of Electrostatic Repulsion

The power of electrostatic repulsion can be very simply demonstrated. Take a length of fine thread and at one end tie a tiny ball of tissue paper. Suspend the thread such that the ball swings freely. Then rub a length of plastic (a plastic ruler well serves the purpose) until it acquires a static electric charge and present it to the ball. Initially, the ball is attracted to the ruler but, as soon as it touches it, it will fly away in the opposite direction. As the ruler is brought toward the ball, the latter will continue to move away, clearly repelled by the electric charge carried by the former.

What is happening here?

At first, the tissue paper ball carries no charge, and is attracted by the charge on the ruler. However, once it touches the ruler, a like charge is transferred from the ruler to the ball and, as like charges repel one another, the two objects now likewise repel. However, because your action on the ruler is stronger than the electrostatic repulsion (you move the ruler in the ball's direction, overcoming their mutual repulsion) it is the ball, being freely suspended, that is obviously repelled.

According to the Steady State theory, matter is being created one atom at a time throughout the universe. This new matter takes the form of hydrogen and it is this gas that fills the vast intergalactic realms. It is also within this diffuse and extremely tenuous cosmic atmosphere of hydrogen that electrostatic repulsion performs its work if there really is a slight difference in charge between protons and electrons. Denser regions—ionized condensations or units—arise within this global hydrogen atmosphere and, since they are ionized, these units are also conducting, resulting in excess charges appearing at their surfaces. These units contract under gravitational attraction and become galaxies and clusters of galaxies, expelling their excess charges in the form of protons. Gas infalling into these units from the general field will have an effective temperature of around one million degrees. The units themselves do not actually repel each other, but they mutually recede because they are swept along by the electrostatic repulsion of the cosmic atmosphere of hydrogen. The rate of creation of this matter and charge is what determines the Hubble constant.

Shortly after the Lyttleton-Bondi paper was published, the results of an experiment by A.M. Miles and T.E. Cranshaw were also published, revealing a proton-electron inequality (if one existed at all) at least a factor of 50 smaller than that required by Lyttletin and Bondi. The Miles-Cranshaw experiment was both elegant and simple. First, they placed a cylinder of compressed gas inside an aluminum box and then placed this box inside a second one, also made of aluminum. The first box was well insulated from the second. Any changes in potential between the two boxes were measured by a sensitive electrometer. The gas in the cylinder was then released to an external container, passing en route through a small gap across which an electric potential was applied in order to remove any ions that may have been present within the gas. After repeating the experiment with argon and nitrogen, no significant charge was found to be carried by the out-flowing gas. From this, the experimenters drew the conclusion that the atoms were effectively neutral and that any inequality of charge that might exist between protons and electrons amounted to less than three parts in 10^{20}. During the succeeding years, repeats of this experiment, as well as further experiments by M. Marinelli and G. Morpurgo using steel balls, have reduced the maximum possible inequality to levels of around one part in 10^{21}. This is entirely consistent with no inequality actually being present.

The disagreements between the early experimental results and the value of the inequality required to yield the results that Lyttleton and Bondi required did not, however, concern these scientists too greatly. In fact, their conclusion would continue to hold even if the inequality was at the 10^{20} level, provided that the usual Maxwell equations are modified. This is possible, they concluded, because strict conservation does not hold in a universe where matter and charge are being continuously created.

The Lyttleton-Bondi model was examined by the third member of the Steady State trio, Fred Hoyle, who found some very interesting and strange possible consequences. Hoyle found that the cosmological constraints placed on the theory (a steady-state solution and modified Maxwell equations) meant that the physical situation existing between two charges would change signs if these were separated by very large distances. In other words, when large separations are involved, two like charges attract and two unlike charges repel; exactly the opposite of what is experienced at smaller separations (and what we were taught at school!)

and which is enshrined in the well-known Coulomb's inverse-square law. Hoyle suggested that this modification of Coulomb's law might even be responsible for the separation of matter and anti-matter in the universe. The separation of particles and anti-particles would, he argued, generate an electric current which in turn would produce an intergalactic magnetic field, further contributing to the sculpturing of the universe at large.

The Lyttleton-Bondi theory, especially as developed by Hoyle, stands as an interesting and thought-provoking alternative to the more conservative cosmologies in which gravity rules supreme. In that sense, it resembles the plasma theories of Alfven and other "electrical" cosmologies, notwithstanding the other very great differences that exist between these theories. Nevertheless, together with these other theories and hypotheses, it came to nothing. The newer, stricter, constraints on any possible inequality of electrical charge between proton and electron demonstrate that, even in the unlikely event that any inequality at all remains, it is far too small to exercise the cosmological effects proposed by these scientists. Furthermore, observational cosmology has moved on since 1959 and the greatest weight of evidence derived from a number of empirical directions clearly supports the evolutionary Big Bang universe over the Steady State model. Direct observations of the Cosmic Microwave Background are totally inconsistent with the latter theory, as is the evidence from changes in galaxy morphologies and the numbers of quasars per unit volume as we look further out into space and backward in time. The latter reveals a clear history of change in the universe at large in a way that is consistent with an evolutionary process extending from some creation point/moment just short of 14 billion years ago. This is, of course, in serious conflict with any theory that requires the universe to have remained in a steady state. Even though some astronomers (most notably Fred Hoyle) made valiant efforts to preserve at least the essentials of the Steady State theory, their attempts became a trifle reminiscent of the Mediaeval models of the Solar System with their introduction of epicenters on epicenters to account for the motion of the planets within a basically Ptolemaic system. Sometimes modifications of a model are valid and succeed, but at other times they reach such a degree of complexity as to become unrealistic and demand that the entire system be surrendered and something new put in its place; the Copernican model in place of the Ptolemaic and the Big Bang in place of the Steady State cosmology. What really requires wisdom however is determining where the dividing line between justifiable modification and unrealistic tinkering should be drawn. Not everyone will agree and it is equally easy to err on the side of conservatism as it is to err on the side of change.

Max Tegmark's Mathematical Universe

Professor Max Tegmark of the Massachusetts Institute of Technology neatly summed up the various attitudes that most of us have toward mathematics as "either … a sadistic form of punishment, or as a bag of tricks for manipulating numbers" (*New Scientist*, 15 Sept. 2007, p. 38). Many school children would agree with the first alternative while the majority of adults other than students of physics or philosophy (and undoubtedly quite a few students of these disciplines as well) would probably, if pressed to think about the subject at all, come out in fair agreement with the second. Mathematics appears to be about manipulating symbols; for instance, much of it seems to deal with strange things called numbers which we unthinkingly identify with marks made on paper or displayed on the keyboard of a computer—symbols that look somewhat, although not quite, like letters. Whether we are children leaning our multiplication tables or post-graduate students presenting dissertations on Hilbert spaces, this is what mathematics seems to be (Fig. 5.1).

Yet, once we begin to think a little more deeply about it, complications with this simple picture soon begin to arise. Thus, although we just said that much mathematics deals with things called numbers and that it is frequently an unexpressed assumption that these are simply symbols, we did not ask what these symbols represent?

Think a little deeper and another problem arises. If by symbol we mean a certain shaped mark and if numbers are simply identifiable with the symbols to which we normally refer when we speak about specific numbers, does that not imply that there is an indefinitely large multiplicity of each number? For example, if the

Fig. 5.1 Professor Max Tegmark (*Courtesy*: M. Tegmark)

number two is simply the same thing as the symbol *2* (or *II* if you prefer the Roman notation) there will be as many numbers two as there are instances of these symbols. So these symbols, however they are written—cannot, after all, be identical with the numbers themselves. On the contrary, they must simply represent or stand for the numbers themselves. On this interpretation, there is just a single number called two, even though there may be countless instances of the symbols *2* and *II*.

But if a number really is something to which the symbolic mark refers and not that symbolic mark itself, then what exactly is it?

We will leave this question dangling for now and move on to yet another issue raised by our deeper thinking about mathematics, namely, how and why does mathematics relate so well to the world around us? Somehow, this method of manipulating symbols opens up insights into the nature of the physical world. Theories of physics, that most basic of the physical sciences, can be—indeed must be—formulated in mathematical terms and it is by the working through of these mathematical expressions that further knowledge is acquired, theories are modified and more details of the nature of the universe revealed. This is so taken for granted by scientists that the wonder of it is most often overlooked. On the face of it, there does not appear to be any a priori reason why mathematics and the nature of the physical universe should be so closely intertwined, yet all evidence shows that such a relationship does indeed exist. The positron or anti-electron, for instance, was first postulated because of a mathematical relation. The square root of an integer has both a positive and a negative solution, so to be true to the reality being described by equations involving square roots, physicist P. Dirac postulated the existence of an anti-electron. The particle was later discovered empirically, but it was first postulated mathematically. It is not intuitively obvious that something as seemingly esoteric as the solution of a square root could predict something in the external world, but that is indeed the situation. Furthermore, we have the example of Werner Heisenberg's derivation of a special calculus by which to describe atomic structure in mathematical form. Unknowingly, Heisenberg had actually reinvented (or, perhaps we should say, rediscovered) something that had already been known as a branch of pure mathematics, namely, matrix algebra! When initially developed, matrix algebra appeared to be purely about arrangements of tables of numbers, yet Heisenberg had (by accident so to speak) discovered that it actually revealed some of the deepest secrets of atomic structure. In the face of examples such as these, who could disagree with the words of physics Nobel Laureate Eugene Wigner when he stated that "the unreasonable effectiveness of mathematics in the natural sciences" demands an explanation?

With a few exceptions such as Wigner (and, as we shall see in due course, Tegmark) physicists, by and large, tend not to worry too much about these sorts of things. On the other hand questions of this nature are what philosophers rely upon for their bread and butter and a number of philosophical theories have been put forward as attempts to explain the mysterious relationship between the world and mathematics. At one end of the spectrum there is the approach stemming from the eighteenth century philosopher I. Kant which in effect proposes a radical distinction between the world as it really is and the world as we perceive it to be. According to the Kantian thesis, what is in truth out there—the universe as it is in itself—is intrinsically unknowable to the human mind. We are forever prevented from perceiving the true nature of reality because we have been born with a certain inner mental filter through which all perceptions of the world must pass and by which they are all in a sense distorted into something that we can comprehend. It is as if we are all forced to see the world through tinted spectacles, although the tints do not refer to color and the spectacles apply to each and every sense, not merely sight. These spectacles or filters—however we might like to picture them—impose upon perceived reality features that the real world does not possess. Even such basic features as space, time and causation are projected onto the world but form no intrinsic feature of it, according to this theory. On this view, the world is describable by mathematics because mathematics forms part of the filter through which we encounter it. In reality however, the world is not mathematical at all. Mathematics may marvelously describe the world as we perceive it and, indeed, the world as it can only be comprehended by us, but it says nothing about the world as it really is. That is as non-mathematical as it is non-perceivable.

A modern reader might find similarities between this hypothesis and the theme of the movie *The Matrix*, although the split between the real and what is experienced is far more fundamental and far more extreme than anything that even *The Matrix* suggests. If the hypothesis is correct, one might indeed wonder if the filters of spectacles are the constant for all species. For instance, if beings not too unlike ourselves exist elsewhere in the universe, would they necessarily possess the same types of mental filters as we possess? Would they experience the world as we experience it, or would their reality be totally different from ours? Would they describe their universe in terms of an entirely alien mathematics (or no mathematics at all!), argue with an entirely alien logic and construct theories based upon an entirely alien physics? If so, we would be completely unable to converse with them, nor they with us. Chances are that both sides would see the other as nothing more than sub-intelligent brutes. Not even experience of each others' technology need be convincing evidence to the contrary, as (being based upon a totally diverse perception of the universe) it is doubtful if each would even recognize the works of the other as technology at all.

We might extend this speculation even further and suggest that the other forms of animal life on this planet have different sets of filters. Do we live in what is effectively a different world to that inhabited by our pet dog or cat?

In the latter case, I think that there is enough evidence to show that the mental filters between humanity and the rest of the animal population are basically the same. It is certainly true that different animals perceive the world differently to those of other species in the sense that some can see parts of the electromagnetic spectrum that others cannot or perceive the flow of time differently to others. Moreover, the degree in which the different senses vary between species alters the perception of the world and changes the degrees of importance between varying aspects of it. Someone once said, for instance, that if dogs could write stories they would all be about smells! Yet, all of these differences are in degree rather than in kind and there is no reason to think that the basic properties of the world are different for dogs as they are for humans. Animals clearly perceive causal connections not too differently to the way in which we perceive them. Hunting relies upon cause and effect operating; puncture the neck of the prey and it will die and therefore be available to eat. The more intelligent animals use basic logic. Scratching at the door at meal time implies a simple process of deduction. "Humans are annoyed by this noise. They know that if I am fed I will stop making it. Therefore if I make the noise they will feed me." An animal probably does not formulate the argument as distinctly as this, but who would disagree that this is not the way it thinks?

As for animals, so most probably for aliens, if any such exist. It is unlikely that we need worry too much about alien logics and alien mathematics, at least not in the radical sense spoken of here. The Kantian dichotomy between the world as it is and the world as we perceive it raises many problems. Although it is reasonable to think that perceptions are filtered to some degree and that this filtering has biological significance, there is no evidence that the world as we perceive it is fundamentally distinct from the world as it really exists. It has been found, for instance, that certain races dependent upon hunting perceive the periphery of their field of vision with greater clarity than the central region. This is because hunters (who also might be the hunted in these situations) must pay more attention to the edges of their visual field than members of urban societies. Yet that in no way alters their fundamental perception of the world. The basic structures of the world remain constant as the framework of reality within which these smaller differences appear.

π in the Sky?

At the other end of the spectrum, we have theories such as that of Plato in which mathematical entities and the whole class of "universals" are transcendent objects of which the observable world is but a shadow. In a sense, these ideas share with Kantian philosophy the opinion that what we experience is not the ultimate reality of the universe, but in every other way Platonist philosophy is diametrically opposed to the Kantian.

The problem which Plato attempts to answer can be conceived in the following way. Back in school, we were taught that "nouns name things". For much of the time, it is easy to see what nouns do actually name. Whether proper nouns of places, common nouns, or collective nouns, the noun points toward some more or less concrete and particular thing. Even the last of the trio refers to something particular, albeit one composed of many parts. But what about the class of abstract nouns, examples of which include such things as "redness", "truth", "goodness" and so forth? Unlike the other varieties of nouns, we cannot actually point to the things that they name. We can, of course, point to red things, true beliefs or good actions, but these are simply particular things which express or instantiate the general or abstract "thing" named by its respective abstract noun. In other words, we may see red things or experience good deeds, but do we see in these redness or goodness themselves; in the abstract so to speak? The nearest we can come to an affirmative answer to this question is to say that we see instances of such generalities in the form of particular things or actions which exemplify these generalities, but we do not experience the generalities, *qua* generalities, themselves. We may even question what it could mean to experience generalities qua generalities, that is to say, to experience them apart from their instantiations in particular entities.

Philosophers call the things named by abstract nouns universals. The problem of universals has been thrashed around by philosophers since the time of Plato and there is still no consensus as to what they are. Outside of the discipline of Philosophy, the problem seldom, if ever, arises and non-philosophers are by and large left wondering what all the fuss is about. At least, that is the situation today. During mediaeval times when philosophical doctrines loomed larger in the minds of the powers-that-be than they do in contemporary culture, ones' doctrine of universals could determine ones' place in the intellectual strata of the day. For instance, William of Occam (after whom the Occam's razor principle of scientific method is named) was excommunicated from the Church because he came down on the wrong side of the universals debate.

This has to do with Tegmark, mathematics and the main topic of this chapter in that the problems raised by the nature of mathematics are really instances of the wider issue of universals and cannot be fully appreciated apart from this broader scene. Returning to Plato, we have already noted that he proposed the hypothesis that understood universals as transcendent entities existing in their own right and, in effect, constituting the ultimate reality of the universe. This position is known in philosophy as Transcendent Realism. The name he gave to universals is sometimes translated into English as Forms and at other times as Ideas. The former term will be used here as the latter tends to give an erroneous notion of what he actually meant. For most of us in today's culture, the word idea carries the sense, made explicit by English philosopher John Locke (1632–1704), of something subjective; something in the mind. But that was not the sense intended by Plato. The Forms were not ideas in this sense at all. Although he believed that they were apprehended by the mind, they existed in a realm outside of the mind, indeed, outside of the world of particular things, minds included. They simply existed, eternal and immutable and would have existed even in the complete absence of both particular things and minds. Plato understood them to be eternal, not in the sense of having existed from an infinite time in the past and persisting for an infinite time in the future, but simply as entities existing apart from time altogether. They just are; and that is that.

Plato's theory did not go unchallenged. His most famous student, Aristotle, developed the counter theory of Immanent Realism which understood universals to be in the nature of essences intrinsic to particulars themselves. This difference of opinion is nicely

Fig. 5.2 Plato and Aristotle as depicted by Raphael (Detail of fresco in the Apostolic Palace, Vatican City. *Courtesy*: Wikimedia)

represented in Raphael's fresco of Plato and Aristotle. Plato, the elder sage, points upward toward the sky while the younger Aristotle, walking beside him, appears to be stressing the universal nature intrinsic to the particular things of the world around us (Fig. 5.2). Of course, Plato did not literally understand the Forms to be up there in the sky and Raphael was not implying that he did. But by having Plato point skyward, the artist is demonstrating that Plato is symbolically demonstrating his opinion that the Forms belong to a higher realm of existence; a realm that, though not up there in the spatial sense, is nevertheless up there in what might be called an ontological sense. In fact, Plato sees his Forms as being not simply universals in the sense of that term as used by later philosophers, but as perfect archetypes of their instantiations within individual things in this lower world of particulars. Within this realm, figuratively though not literally in the sky, mathematical entities (as universals) also have their home. Up there is, for instance, the Triangle—the perfect archetypal Triangle of which all individual triangles, all instances of triangularity, in the world of particulars are but imperfect copies. The Platonic world of Forms therefore can be called a realm of mathematical entities, although it is more than that insofar as not all universals are mathematical entities. But, as we shall see, to the extent that the world of Forms is mathematical, Plato and Tegmark find common ground.

In Plato's opinion, whereas particular things in the familiar world are perceived by the bodily senses, the Forms are directly apprehended by the intellect or understanding. There is an amusing story of a conversation about this very issue between Plato and one of his erstwhile students, Diogenes the Dog—the same one who carried a lighted lantern through Athens in broad daylight in search of an honest man. Diogenes criticized Plato's theory by pointing out that, whereas particulars are readily perceived, the Forms are not. How therefore, can these latter entities be afforded such a pivotal role if their presence lacks observational evidence? Plato responded that while particulars are known through the physical senses, the Forms can only be known through the intelligence, and that is the reason why Diogenes failed to know them.

Theories about realms existing in some sense alongside our own may seem strange and esoteric to many people in our day and age, yet it is interesting to note how many times similar ideas crop up in other cultures, some of them not very far removed from our own. The German protestant mystic Jacob Boehme (1574–1624), for example, taught the existence of an Eternal Nature of which the material world is like a dark shadow. Boehme's influence was widespread and his ideas can be found in the poets Milton and Blake as well as, in more recent times, in the writings (at least in the fantasy writings) of C.S. Lewis. But thousands of years before any of these writers, or before Plato for that matter, beliefs on the same continuum of ideas were held by members of the world's oldest living culture, the aboriginal Koori people of Australia. The concept of the Dreaming, as understood by the Koori themselves, has certain similarities with Plato. Unfortunately the English translation as Dreamtime has obscured this fact, instead giving the impression of some sort of mythical golden age of long, long ago.

But the concept, as the Koori hold it, is more complex than this. As explained by Silas Roberts, Elder of the Yolnhu tribe

> By Dreaming we mean the belief that, long ago, these creatures [the mythological inhabitants of the Dreaming] started human society. These creatures, these great creatures, *are just as much alive today as they were in the beginning. They are everlasting and will never die. They are always part of the land and nature, as we are.* Our connection to all things natural is spiritual. [Emphasis mine]. (Quoted from *Australia, The Story of Us* Issue 1).

There are differences of course, but there is also a thread connecting the Dreaming, Plato's realm of Forms, Boehme's eternal nature and other beliefs of this type. This does not mean that such ideas are true, but it does imply that the culture that rejects them (our culture by and large) is the one that is out of step. This is worth keeping in mind as we come to look at Tegmark's theory of a universe of mathematical structures. That is not to equate these ideas with the Dreaming of course, but it may just be true that a Koori person educated in modern cosmology might come to a more ready appreciation of Tegmark's position than many of his contemporary European readers.

The Universe = Mathematics

Turning now from these broader issues, let's focus more specifically on this theory put forward by Tegmark, a concise popular level presentation of which is given in the journal *New Scientist* for 15 September, 2007, and a more detailed account in his 2014 book *Our Mathematical Universe*.

Tegmark begins by drawing attention to the fact that mathematics has played, and continues to play, a vital role in our growing understanding of the physical universe and takes this as strong evidence that there is some essential association between physical science, and therefore the universe that it describes, and mathematics. He approvingly quotes Galileo's statement that the universe is a grand book written in the language of mathematics. Thus far, he has not expressed anything that would arouse controversy; not at least amongst physical scientists. But he is only at the beginning of his argument! In his own words he is about to "push this idea to its extreme and argue that our universe is not just described by mathematics—it is mathematics". Our physical reality is a mathematical structure. In other words, we all live within a gigantic mathematical object! He further argues that while such a position might sound farfetched (even using the word weird in places) it nevertheless can be used to make predictions about the structure of the universe and even holds promise for narrowing down the contenders for an ultimate Theory of Everything. The predictions that it makes concerning the structure of the universe are indeed startling, as will be seen in due course.

Tegmark lays the foundations of his argument by making the assumption that the universe exists independent of human beings. In the minds of the majority of folk, that might seem an assumption too obviously correct to even need expression. Remarkably however, it is actually rejected by the school of philosophy known as solipsism as well as by certain subjectivist interpretations of quantum physics. Adherents of these views reject the belief that we can have viable knowledge of the external, objective, world at that, at least in theory, objective reality cannot be shown to exist. Nevertheless, they cannot do other than accept it for all practical purposes (which surely implies something about the inadequacy of their theories). This division between what one theoretically asserts and what one assumes to be true in the practical sense was aptly exemplified by one solipsist who complained that there were not more solipsists.

Nevertheless, despite such skeptics we may, I think, be pretty confident in accepting Tegmark's first assumption. There is a world out there that does not depend upon either our presence or our observation of it and it is the basic task of physics to describe this world. As Galileo implied, the ultimate description of this objective environment will be mathematical in form.

At this point, Tegmark's original assertion enters in the form of a second assumption, namely, that this external world—the universe—is not only mathematical but is mathematics itself. Physical reality is, in Tegmark's words, a mathematical structure. This assumption is far more controversial than the first and at face value might even appear absurd. So what exactly does this startling statement really mean?

Reviewing Tegmark's book, *Our Mathematical Universe*, in *The Wall Street Journal*, Peter Woit makes the perceptive point that there is a certain slipperiness or ambiguity in that little word is. Stating reality is a mathematical structure could mean no more than the uncontroversial position that reality is structured in such a way as to be describable by mathematics. Galileo would have agreed with this, as would just about every physicist who ever lived. This uses "is" in a soft sense which is more or less definable as "can be described by". But it quickly becomes apparent that Tegmark is imbuing the little word with a far harder meaning. He is giving it the sense of the "is" of identity. And this "is" works both ways, like the equals sign in an equation: "the universe is mathematics" and, equally, "mathematics is the universe".

Tegmark states that, at present, even our most successful theories of physics describe only parts of reality; only certain aspects of the universe. The two most successful theories of them all, general relativity and quantum mechanics give good accounts of gravity and the universe as it exists on large scales in the case of general relativity and of the behavior of subatomic particles in the case of quantum theory. But what he calls "the holy grail of theoretical physics"—the Theory of Everything—would present a complete description of the whole of physical reality encompassing alike the cosmic and subatomic scales. Tegmark asks what such a theory would be like and it is in answering this inquiry that the force of his assertion that reality is mathematics becomes apparent. "If", he argues, "we assume that reality exists independently of humans, then for a description to be complete, it must also be well defined according to non-human entities—aliens or supercomputers, say—that lack any understanding of human concepts" (*NS*, p. 38). Human concepts are, he says, baggage of which a Theory of Everything must be stripped before it can adequately describe a universe that is essentially independent of human beings. The baggage that must go includes all of those concepts which can be named by words such as particle, observation, star, atom etc. In other words, all that constitutes the concrete world that is the subject of human observation and inquiry.

All physical theories thus far postulated have, according to Tegmark, two components; mathematical equations and words that explain how these equations are connected to what we observe and intuitively understand. It is these words that name the things of the world and are, as Tegmark expresses it, concepts created by humans. But that is exactly what he previously named as baggage. In principle, he argues, everything could be calculated sans this baggage. "A sufficiently powerful supercomputer could calculate how the state of the universe evolves over time without interpreting it in human terms." That is what the Theory of Everything would look like; a description of external reality involving no baggage—nothing introduced by human beings. Such a theory would offer a "description of objects in this reality and the relations between them [that would be] completely abstract, forcing any words or symbols to be mere labels with no preconceived meanings whatsoever. Instead, the only properties of these entities would be those embodied by the relations between them" (NS, p.38).

Now, a "set of abstract entities with relations between them" is precisely what a mathematical structure is. In the final analysis therefore, reality—the universe—is a mathematical structure. It is mathematics in the most radical sense, not merely something that is amenable to mathematical description and analysis.

Even the symbols by which math is expressed are part of the baggage, not aspects of the mathematical structures themselves. They are "mere labels without intrinsic meaning". Whether we write two plus two equals four, dos mas dos igual a cuatro, $2+2=4$, $II+II=IV$ or any other equivalent notation, we are only denoting the entities and the relations between them. Therefore, any notation is, in the final analysis, nothing more than irrelevant baggage. All that really exists are the integers and the relations between them—and it is the relations between integers that constitute the only properties of the former. These mathematical structures are not invented by us; mathematicians discover them, they do not create them. According to Tegmark, "[they] are not 'created' and don't exist 'somewhere'—they just exist" (p. 41). Such a statement could almost have come from the mouth of Plato himself (note also the similarity between Tegmark's assertion and the emphasized portion of the earlier quote by Elder Silas Roberts concerning the creatures of the Koori Dreaming). These mathematical structures exist in their own right as abstract immutable entities outside of space and time according to Tegmark's radical Platonism (his own terminology). According to his interpretation, the "mathematical structures in Plato's realm of [Forms] ... exist 'out there' in the physical sense".

Frog Perspectives and Bird Perspectives

Ironically, if Tegmark's line of reasoning is correct, the closer we approach a theory that yields a complete explanation of what the universe really is and how it operates, the further we seem to be getting from the observable familiar world of our everyday experience. If such a theory ever does explain reality, we might be excused for thinking that what *is* finally explained is not our familiar reality at all. Following Tegmark, it is almost as if some version of Plato's world of Forms is the real subject of the long-sought Theory of Everything, whereas our continued experience is of the shadow world of particular entities. We might be excused for having sympathy with Diogenes and worry that Plato's rather caustic reply to his criticism might also be directed toward us.

A possible way out of this confusion is supplied by Tegmark with his suggestion that external reality may be better understood if we recognize two different, albeit complementary, ways of viewing it. One of these ways he calls the outside overview of the physicist studying the mathematical structure of the universe and which he compares to that of a bird in flight, surveying a landscape from high above. The other perspective is that of an observer living inside the universe of familiar experience, which he compares to the view of a frog living within the same landscape as that being surveyed from above by the bird. Needless to say, the perspectives of these two creatures yield very different views of the world.

One important difference between these two perspectives involves the perception of time. From the bird's-eye view of the physicist, what the frog perceives as something moving with a constant velocity will be seen as "a straight strand of uncooked spaghetti", as Tegmark picturesquely expresses it. Furthermore "Where the frog sees the moon orbit the Earth, the bird sees two intertwined spaghetti strands. To the frog, the world is described by Newton's laws of motion and gravitation. To the bird, the world is the geometry of the pasta."

Another difference in the two perspectives concerns the amount of information required to fully describe each view of the world. The frog's perspective would need a truly colossal amount of information for a complete description. It is estimated that a truly complete description of the empirical world, down to the position of every grain of sand and every star in the sky, would need something like 10^{100} information bits. By contrast, a complete description of the bird's-eye landscape—that is to say, a true and complete Theory of Everything—would require (most physicists believe and hope for) only a small amount of information; something small enough, as Tegmark expresses it, to fit in a book, if indeed not on the back of a T-shirt.

It is not stated that the bird's-eye perspective is somehow more real than the frog's-eye one, but Tegmark's implication is that it is in a true sense the more basic or fundamental one. The bird sees the nature of reality more clearly than the frog sees it. We are tempted to express this in Platonic terms by saying that the bird perceives the Forms whereas the frog only perceives the particulars. The bird is Plato and the frog, Diogenes.

Yet, despite the similarities in the theories of Plato and Tegmark, the latter is first and foremost a physicist and not a philosopher. To present his theory as one of physics and not of metaphysics, he must show that it makes specific predictions that can either be verified or falsified by further observation.

The first prediction arising from the mathematical universe hypothesis is that further mathematical regularities will be discovered in nature as science advances. This prediction is being verified all the time. It is precisely this phenomenon that gave rise to the statements of Galileo and Wigner. Yet, I find it difficult to accept that this prediction follows uniquely from the mathematical universe theory as Tegmark conceives it; from what we might call the strong form of the theory. Surely, this prediction also follows from weaker forms of what we might still like to term the mathematical universe hypothesis, namely, the hypothesis that simply holds reality to be describable in mathematical terms or states that in some manner it includes mathematical structures, without being identical to them (more will be said about this alternative later in the present chapter). The fact that mathematics leads us to discover deeper truths about the nature of physical reality and the corresponding fact that each new physical discovery leads us to other levels of mathematical expression does indeed constitute good evidence that the universe is mathematical, and to this extent we fully agree with Tegmark.

Nevertheless, we do not agree that this necessarily supports the mathematical universe hypothesis in the restricted sense in which he uses the term. It is certainly consistent with this, but being consistent with something only goes part of the way in the journey of verification.

His second prediction is much more radical and, we might add, vastly more controversial as well. He asserts that the hypothesis predicts the existence of parallel universes. Indeed, he sees the existence of a multiverse of alternate universes as an inevitability if his version of the mathematical universe hypothesis is correct. The theory gives no other option but to assert the existence of other universes. But why should that be so? Why should a theory of *our* universe imply the existence of a multiplicity of other universes? The reason, Tegmark argues, is that the mathematical universe theory is by definition a complete description of reality. He states, "If it lacks enough bits to completely specify our universe, then it must instead describe all possible combinations of stars, sand grains and such—so that the extra bits that describe our universe simply encode which universe we are in, like a multiversal phone number" (NS, p. 40). In that way, it is simpler to describe a multiverse than it is to describe just one universe. Moreover, this hypothesis describes a multiverse of a special and, we might say, extreme or even weird variety, but to better appreciate this, we will need to take a look at the different types or levels of multiverse as differentiated by Tegmark.

A Universe of Universes?

The concept of a multiverse or a multiple universe has been around for quite a number of years. In fact, there is not just a single model of the multiverse. Quite the contrary, several models exist and it is even possible that all of them are correct—or, equally, that all of them are incorrect and the universe is single after all. Recognizing this, Tegmark divides the models into four classes or levels as already mentioned. The different levels are as follows:

Level I. This is the simplest multiverse model and the one dependent upon the least number of assumptions. It appears to follow naturally from any cosmology that predicts a universe infinite in spatial extent and follows from the fact that the velocity of light is finite. As just about every book on elementary astronomy informs its readers, because of this finite velocity, the further we look out into space, the further we also look back in time. Now, as most cosmologists today accept some version of the so-called Big Bang model and as this event is now estimated to have taken place just a little short of 14 billion years ago, there is effectively an absolute cosmic horizon beyond which we cannot see. We cannot see beyond the Big Bang, because if this event really was the creation of the universe—the creation, not just of matter and space, but even of time itself—there simply is no beyond and no before. We cannot even accurately say that there is nothing beyond/before the Big Bang, as this very phraseology subtly implies that something (a great emptiness if nothing else) is extant there. But even that is not radical enough. Before the creation of all physical reality, not even a great void could exist. There would simply be no physical existence at all. Indeed, there would not even be any time in which any such void could exist.

Because the Big Bang happened almost 14 billion years ago, it might be thought that the ultimate cosmic horizon lies at a distance of 14 billion light years. That, however, fails to take on board an interesting if rather weird fact about the universe. We know that it is expanding on the large scale. Space within the Solar System, or even within our galaxy of cluster of galaxies is not expanding, but on cosmic scales the galaxy clusters are racing away from one another like painted dots on the surface of an inflating balloon. Yet, this cosmic recession does not imply that the galaxies and clusters of galaxies are racing away from each other through space. What it actually means is that space itself is being stretched. In effect, there is a continuous creation of space within the universe. There is more space in existence today than there was yesterday. One result of this is that so much space has been created since the Big Bang that instead of our view of the universe being restricted to 14 billion light-years, we can actually see out into space for something closer to 46 billion light-years. Or, put another way, the furthest reaches of the universe within our sphere of observation are not 14, but 46 billion light years distant.

This accessible portion of the universe is sometimes termed the O-sphere or observable sphere. As far as we are concerned, it is our entire universe. However, if the actual universe is significantly larger than the observable universe, i.e. the O-sphere, there must be other O-spheres beyond our own, some of which (the nearer ones) will actually overlap our own. If the actual universe is truly infinite, the number of O-spheres will also be infinite. In effect, each O-sphere will be another (observable) universe, although there is no reason to think that these will differ in terms of their fundamental physics from our own observable universe. On the contrary, as they all arose from the same creation event, there is every reason to suppose that the basic laws of physics remain constant throughout the actual universe and, therefore, between all the potentially observable universes.

The ensemble of alternate observable universes (or potentially observable universes as there is no a priori reason to think that intelligent life will be common to them all, although that assumption is usually made) constitute Tegmark's Level 1 multiverse. However, because each O-sphere is finite in extent and contains a finite number of particles, their variety will be finite, even if their number is not. Tegmark estimates that even if every O-sphere was filled with particles, the number of possible combinations of these is the extremely large, albeit still finite, 2 to the power of 10^{118}. Forgetting about the restrictions that Special Relativity places on superluminal flight, let us imagine that we have procured a spaceship capable of near infinite velocities and that we have decided to journey through O-sphere after O-sphere and note the entire variety of the actual universe. The first thing that we would notice about the nearer O-spheres is that they all look basically similar to our own on large and medium scales. That is to say, there will be galaxies, stars, planets and so forth just as there are in our home universe. We may even encounter other forms of intelligent life and at least some of these might look superficially like human beings. But there would be no specific objects with which we are familiar. But as we travel further away from our home O-sphere, we will find that clones of objects familiar in our observable universe begin to appear and if we continue to journey for the amazing distance, which Tegmark calculates to be about of 10 to the power of 10^{118} m from our starting point, we would arrive back home! We would arrive, that is to say, in an O-sphere that was in every respect identical to the one we left. That O-sphere would even have counterparts of us, however we would be unable to meet them. They would have left in their spaceship, at the exact time that we did, to travel at near infinite velocity through the universe of O-spheres. Indeed, the family and friends who greet us at our "return" would welcome our arrival back home, just as our family and friends are doing for our counterparts in the O-sphere from which we initially departed!

In *Weird Universe*, it is argued that as there is no way of distinguishing two identical O-spheres, we might as well consider them to be literally the same, ending with the paradox that the infinite universe is actually finite! No more will be said about this here however.

There is one further point concerning the existence of Level 1 multiverses though, and that is that even if the actual universe is not literally infinite, O-spheres which are to a greater or lesser degree copies of our own will occur if its extent is nevertheless extremely large. As we have just seen, if the universe is equal to or greater than 10 to the power of 10^{118} m in extent, exact copies of the entire O-sphere will appear. For there to be an O-sphere containing an exact copy of yourself, Tegmark calculates an extent of 10 to the power of 10^{29} m would be required.

Level II. According to inflationary theory, our observable universe and much of the actual universe beyond began as a bubble which inflated enormously over an extremely brief period of time. The properties of the universe as we know it depended vitally upon both the rate of this inflationary epoch and the length of time over which the inflation continued. Some versions of the theory speculate that other bubbles were also generated and, if that is true, there is no obvious reason why each should be expected to expand at the same rate or for the length of time taken by the expansion epoch to have been constant for all. Accordingly some, though not all, of the constants of physics would vary between the different bubbles. Many bubbles would probably acquire physical properties so extreme that they would either collapse into black holes or, at the other extreme, inflate into effectively empty spaces, but others would presumably evolve into universes having some characteristics similar to our own albeit differing to greater or lesser degrees in the finer physical details. Unlike the O-spheres of Level 1 however, not even a hypothetical superluminal spaceship could travel to these universes. They are cut off completely from our own.

Level III. Multiverses of this level are implied by the interesting interpretation of quantum mechanics given by H. Everett in the 1950s. A problem arises in quantum theory in so far as the quantum wave function predicts many different states, only one of which is relevant to the observable universe. This is further complicated by the debate as to whether the wave function is somehow real or little more than a convenient way of speaking. If the former, does this imply that all of the states are equally real and, if so, how is this puzzle going to be addressed? Everett put forward the view that the wave function is indeed real and that each predicted state is also real, although only one will actually be observable to us. The universe model that he proposed has been likened to a garden having branching paths, with the universe splitting into a separate branch (in effect, an alternative universe) at every point where more than a single possible outcome occurs. Everett even extended this to a belief in quantum immortality; the hypothesis that we are always alive in some alternative branch of the universe, even after we are dead and buried in this one.

Everett's theory was not exactly welcomed with open arms by the quantum hierarchy at the Niels Bohr Institute in Copenhagen, but did win an early supporter in John Wheeler. Wheeler's contribution actually led to the theory sometimes being given the title of the Everett/Wheeler interpretation of quantum physics. Nevertheless, Wheeler later abandoned the theory as a viable interpretation, arguing that it was saddled with too much metaphysical baggage to be satisfactory. Not everyone agrees with that assessment however, and the theory has become more popular in recent years, although it is still not the choice of the majority of quantum physicists.

In fact, contrary to what statements by their supporters may at times imply, all of these multiverse theories remain controversial. More is said about this in *Weird Universe*, but for the present purpose it is enough to say that the very bases of these theories have been challenged and cannot be assumed as proven facts. The infinite extent of the universe, for instance, has been challenged by topologist Janna Levin and colleagues who argue that even if the geometry of space is precisely Euclidean, that does not necessarily imply that it is of infinite, or for that matter even of very large, extent. The shape of the universe could, for example, be that of a cube and the Euclidean space that we see is simply one of its (finite) faces.

With reference to the Level III multiverse models stemming from Everett's interpretation of quantum physics, Heinz Pagels in his book *The Cosmic Code* argued against the popular understanding of this interpretation, namely that all macroscopic possibilities are fulfilled in innumerable alternative branches of the universe. Pagels argues that, even if the Everett interpretation is accepted for the sake of argument (and he does not accept it in actual fact), it does not necessarily follow that these so-called alternative universes are truly distinguishable from one another at macroscopic scales. On the contrary, Pagels suggests that they might be related to each other in a manner similar to that of molecules in a portion of gas. Although the gas molecules in, say, a jar assume an enormous number of alternative configurations, when the bottle of gas is viewed as a whole, all of these many configurations are indistinguishable. Maybe, even if the controversial interpretation of Everett is accepted, the alternate configurations are only apparent on the quantum level. In a sense, there may be innumerable alternate quantum worlds incorporated in a single macroscopic reality.

Level IV. The various levels discussed above do not, however, have any necessary bearing upon the mathematical universe hypothesis. The level of multiverse predicted by this hypothesis is one in which all possible mathematical structures are instantiated in some universe in the same way that the mathematical structure which constitutes our universe, and whose properties correspond to its physical laws, are instantiated therein. Each alternative mathematical structure defines a universe within this ensemble of mathematical universes. Moreover, as Tegmark assumes that these mathematical structures simply and necessarily exist, this hypothesis, if correct, makes the existence of such a multiverse—a Level IV multiverse—compulsory.

Tegmark sees in this type of multiverse the solution to the problem that is often posed as to why our universe has a certain set of parameters, describable by one set of equations, and not by others. If we live in a level IV multiverse however, the alternative parameters and alternative equations are indeed relevant somewhere—albeit not in our universe. In short, the different universes

in a level IV multiverse are describable by different types of physics. Perhaps some of these (maybe even most of these) alternative
universes have properties that do not allow for life to be present.
Yet, because we are dealing with objective reality, these are no less
real than ours, even if they are not the objects of perception by
sentient physical beings.

The level IV multiverse has sometimes been called an anything goes model in the sense that if all possibilities are realized in
some universe, then every imaginable universe must exist within
the ensemble that constitutes the multiverse. On this assumption,
there must be universes where even the wildest inventions of fantasy are real. Does this imply that there are actual alternate realities
where Alice really does go down the rabbit hole into Wonderland
and Terry Pratchett's *Discworld* really exists? To questions of this
nature, Tegmark answers with a resounding no. We can, he states,
imagine all kinds of things that are mathematically undefined.
Such imaginary worlds do not correspond to mathematical structures and therefore cannot exist—or at the very least, need not
exist—in the level IV multiverse as he conceives it.

Although the similarity cannot be pushed too far, there is a
certain commonality of argument between the prediction of a level
IV multiverse from the reality of mathematical structures and that
of a level III universe from the hypothesis that the quantum wave
function is physically real. In each instance, once either mathematical structures or the wave function is hypothesized to have a real
existence out there in the objective world, it follows that the different predicted states, in the one instance, and the different predicted
mathematical structures in the other must in some sense be just as
"real" as the state and structure of our own universe. Multiple universes appear to follow as an inevitable consequence. However in
the case of the predicted level IV multiverse, Pagels' criticism of the
usual interpretation of Everett's level III will not hold. Tegmark's
insistence that the position for which he is arguing concerning the
essential mathematical nature of the universe necessarily predicts
a multiverse of the fourth level appears to be inescapable, no matter
how counter-intuitive and difficult to imagine it may be.

That, however, does not worry Tegmark. Quite the contrary
he feels that the counter-intuitive nature of the hypothesis may
actually be part of its strength. Darwinian evolution, he argues, has

"endowed us with intuition only for those aspects of physics that had survival value for our distant ancestors" implying that other aspects of physics that have little or no survival value are likely to appear counter-intuitive or even downright weird to us. As examples of such counter-intuitive aspects, he names the slowing of time at relativistic speeds and the changes of identity of colliding sub-atomic particles at high temperatures. He admits that, for him, "an electron colliding with a positron and turning into a Z-boson feels about as intuitive as two colliding cars turning into a cruise ship" (*NS*, p. 41).

While a strict adherence to Darwinian evolution would appear to restrict our intuition to a narrow field of physics, it seems to me that it would equally restrict our curiosity. Years ago, Aldous Huxley used the term "biological luxuries" to denote certain features of the human mental makeup that seemed to go beyond a strictly survivalist explanation. Huxley was thinking in particular of our sense of color, which seems to go far beyond what is necessary for survival but which nevertheless is responsible for enriching our lives through artistic endeavor as well as through the general enjoyment of this spectacular aspect of the natural world. But it may, I think, be argued that the sense of wonder and the urge for inquiry into the nature of that surrounding world is another biological luxury which we possess. The development of physics may then be seen as the means whereby we can probe the deepest questions of physical existence and thereby hope to satiate this curiosity. On the face of it, this cosmic curiosity, as we might term it, could be seen as injurious to survival. The story told of both of the ancient Greek sage Thales of Miletus and of French astronomer Charles Messier falling over an obstacle whilst contemplating the stars exemplifies the point, even though it may be apocryphal in both instances. Survival involves being very aware of one's immediate environment and not becoming distracted by cosmic curiosity about the nature of the universe. But it is precisely this cosmic curiosity that has taken us into fields where our intuition breaks down, yet there is also an intuitive sense that the whole thing connects together and that even the least intuitively obvious of theories must, if they are to be accepted, at some level explain our familiar world. This is, ultimately, the acid test as to whether a theory will or will not be acceptable as an advance in our description of reality.

But Is the Universe Really Mathematics—Or Is It Merely Mathematical?

The philosophy of logic was taught by Lithuanian/Australian philosopher Professor Vytautas Vaclovas ("Bill") Doniela. Doniela had studied under the influential philosopher John Anderson at the University of Sydney during the 1950s. Anderson argued for an objective realist philosophy in which physical reality—the universe describable by physics—is understood as consisting of material objects locatable within a spatio-temporal frame of reference. Going against the philosophical tide of the time, he was less than enthusiastic about the growing interest in symbolic and mathematical logic, which tended to be understood by its enthusiasts as a pure calculus. Instead, he developed the traditional logic of Aristotle, which he saw as having a stronger foundation in the objective world and which, as a consequence of this, was more in harmony with a realist ontology. Following Aristotle's own conception of logic, Anderson saw this as descriptive of the world itself, or of certain structures within the external world and not simply as a set of laws of thought by which arguments concerning that world should be conducted. He stressed the propositional theory of truth, which for Anderson was tantamount to the propositional model of reality. Logical propositions are not merely statements but structures present within reality itself which the statements about this reality merely reflect. It was this ontological foundation of logic that interested Doniela and was further developed by him in his doctoral dissertation, on the relationship of logic to ontology, at the University of Freiburg and, somewhat later, during his Newcastle lectures. Mathematics interested him less, but because logic and mathematics are so closely related, questions concerning the ontological status of mathematical entities such as numbers were not long in raising their head in his lectures. These issues held a great fascination for the present writer and were taken up in further seminars and discussions (Fig. 5.3).

It is worthwhile to give a brief overview of the position taken by Doniela as it provides both a comparison and a contrast with the mathematical universe hypothesis of Tegmark. Doniela argued that both logic and mathematics describe structures discovered within an objective, physical and spatio-temporal universe. Logical

FIG. 5.3 Emeritus Professor Vytautas V. Doniela (*Courtesy*: Sydney Lithuanian Website SLIC)

propositions are one such class of structure, numbers and mathematical entities in general are others. But what sort of structures are they?

Referring to logic, Doniela's principal interest, he argued that propositions referred to what he called "non-temporal spatial cross sections" of aggregates of things. Take, for instance, the simple syllogism known to traditional logicians as the AAA or "Barbara" (bArbArA) syllogism. This consists of three so-called A-type propositions, that is to say, three propositions of the type "All X are Y" or, to use standard symbolism "X a Y". Two of these constitute the premises and one the conclusion, for example;

All humans are mortal. (H a M).
All politicians are human. (P a H).
Therefore all politicians are mortal. (Therefore P a M).

Superficially, this simple argument might appear as an exercise in thought, but a moment's reflection tells us that it is not primarily about thinking at all, despite the title "laws of thought" often given to logic in discussions of the subject. Nor is it about the language we use to express this thinking process, as some philosophers have argued. Rather, it is about a situation occurring in the objective world; the world of human beings, politicians and organisms with limited lifetimes. In Doniela's theory, what we effectively have here is a relationship between agglomerations of individuals. There is implied an agglomeration of mortal organisms which includes and exhausts a more limited agglomeration of human beings which in turn includes and also exhausts a very much more limited agglomeration of individuals called politicians. By includes and exhausts it is meant that there is no section of the agglomeration of humans which does not lie within the broader agglomeration of mortal organisms and no section of the smaller agglomeration of politicians that does not lie within the agglomeration of human beings. These agglomerations are scattered in space and time across the face of the planet. But the syllogistic relationship exists as an instantaneous snapshot of the relationship between these agglomerations. With Tegmark in mind, we might say that it takes a bird's-eye view of a section or cross-cut of the spaghetti-like world tubes of each of these agglomerations. Although it is difficult to see what changes in temporal extension might in practice alter in this instance, it is theoretically possible that some day an indestructible robot might be elected to Parliament or Congress somewhere in the world. That would cause one of the premises, as well as the conclusion, to change but the logical relationship of the AAA syllogism, or of syllogisms in general, would not by that means be altered.

The assertion that logic is primarily about the world rather than thought processes or language becomes more apparent when syllogisms are represented pictorially in the manner of Fig. 5.4. The syllogism depicted there is the following;

All flowers are living organisms (F a Lo)
Some flowers are red objects (F i Ro)
Therefore, some living organisms are red objects (Therefore Lo i Ro)

This syllogism is represented pictorially and geometrically in the illustration. The agglomerations (termed classes in traditional logic and sets in mathematics) of flowers, red entities and living organisms, though scattered in the actual universe, are depicted here in the form of intersecting circles. The intersecting region shown in red is the conclusion of the syllogism. It is where the classes of living organism, red objects and flowers overlap. In reality, of course, there are red living organisms that are not flowers,

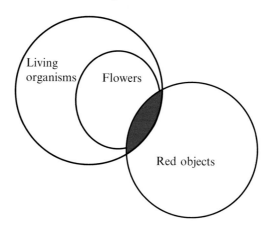

FIG. 5.4 Diagrammatic representation of syllogism (*Courtesy*: The author)

but that is not relevant to the syllogism being considered here. It can be considered a further discovery about the world and its contents. Incidentally, the intersecting portion of any two classes is technically known as the lens—for obvious reasons. The portion of each class outside the lens is known as the lune because of its similarity, as represented in these circular diagrams, to the Moon in various phases (somewhat like a partially eclipsed Moon in the present example).

Coming to mathematical relationships, Doniela takes a similar stance in so far as he argues that these also exist between non-temporal snapshots of agglomerations of things. For example, looking at the very simple equation "$2+2=4$" what we have amounts to a statement that an agglomeration consisting of a thing and another thing is identical to an agglomeration of a thing, a thing, another thing and a further thing. Like a logical syllogism however, this is an instantaneous view or snapshot (by Tegmark's bird, we might say) of the situation. The relation denoted by the plus sign has no extension through time. It is not to be confused with the action of physically adding things together, even though we might loosely use the word that way in our everyday talk. In this latter (we might say temporally extended) sense of add, two things added to a further two things may not result in four things at all. If the objects added together are shards of glass and the addition is violent enough, we might end up with, not four but 40 shards of glass. Similarly, if we were to add two drops of water to another two drops, at the end of the process we still would have only two (albeit somewhat larger) drops of water. Yet, nobody seriously argues that these examples, and innumerable others that we could think of, violate the validity of $2+2=4$. We can have such confidence because the type of physical addition or temporally-extended addition (for want of better terms) is a process in time whereas the true mathematical relation of addition is not. Of course, time enters the scene in some circumstances such as, for example, when we add up the total number of English battles since the year 1066. But here time is seen in the manner of a spatial dimension rather than a process; as seen through the eyes of Tegmark's bird rather than through the eyes of his frog. An agglomeration of events spread out in time is treated no differently from an agglomeration of things spread out in space in the sense that the entirety of the agglomeration is treated as if it were instantaneously present, not coming into existence through temporal process.

It might be worth mentioning at this juncture that, although nobody appears to have argued against the validity of mathematical equations by confusing the two abovementioned senses of add (or, for that matter subtract, divide and so forth) an argument proposed by one philosopher may have depended upon a similar error with respect to logic. In the 1963 edition of his book *Metaphysics*, Richard Taylor argued for fatalism on the ground that the basic logical law of excluded middle (either X or non-X, where non-X is the logical opposite of X) would be violated if a future state of affairs was indeterminate. The logical opposite of a thing includes everything in the universe that is not the thing in question. To take an example, although most people would say that black is the opposite of white, logicians reply that black and white are only popular opposites. The true logical opposite of white is non-white. This includes black of course, but it includes much else besides. In fact, it includes everything in the universe that is not the color white. Taylor argued that the radical X or non-X (white or non-white for example) dichotomy remains true irrespective of the passage of time. Nothing in the future can be somewhere in between X and non-X; between white and non-white for instance. Without going into the detail of Taylor's argument here, what he concluded was that either the laws of logic apply to the relationship between the present and the future or the passage of time somehow causes the law of excluded middle to be violated. If that is true, it opens the possibility that all logical laws are violated by the passage of time; a position that he found disturbing. Doniela did not deal with this issue in particular, but at the risk of putting words into his mouth, he would surely say that the law of excluded middle, like the rules of syllogism or the rules of addition, only applies to an instantaneous snapshot of the situation. The law of excluded middle really means, in his opinion, that at this precise moment, the universe is strictly divided between X and non-X. He would argue that this law, as a law of logic, is not so much violated by the passage of time as irrelevant to it. In short, to argue in favor of fatalism in that way is on a par with arguing that $2+2=10$ because that's how many shards of glass I ended up with after smashing two pairs of them together. We might see this example as analogous to X changing into non-X over a period of time. In any snapshot during this process, the law of excluded middle is valid. Yet, the changing nature of X is, in a sort of way, something that is neither X nor non-X, albeit only if the passage of time is added to the mix. It appears that the frog and the bird perspectives are being confused here. We suggest that Taylor's fatalistic conclusions, like $2+2=10$, come from a confusion of these two perspectives. Much more could be said about this matter but, interesting though that may be, it would lead us too far from the principal purpose of this discussion.

The above is, very briefly, the position held by Doniela concerning the manner in which logic and mathematics relate to the objective world. In common with Tegmark, he saw the world as something existing apart from and independent of the human mind. In agreement with Tegmark, he located mathematics and logic within the structure of that external world and not within the thought processes of the human mind. The query posed by Wigner as to why mathematics should be so unreasonably useful to the physical sciences would have been answered similarly by both Tegmark and Doniela. For each, the universe is mathematical and each would feel quite at ease in predicting that the further advance of physics will continue to find ever deeper mathematical structures in the description of the universe.

Stated thus, it might appear that the respective positions of Doniela and Tegmark lie next to one another on the philosophical spectrum. Yet that would be an illusion. Dig a little deeper and the two positions separate, each to the opposite ends of that same spectrum.

The basic difference, the really large gulf that separates these two positions, lies with the manner in which each of these thinkers understood what we might term the ontological priority of mathematical structures. Tegmark saw these as prior (in this ontological sense) to the universe of things, whereas for Doniela, it was the other way around. For Tegmark, the ontological priority of mathematical structures, in effect, imposed these upon the physical universe, whereas Doniela understood them as purely empirical entities emerging from the actual structure of the physical universe itself. Unlike Tegmark, Doniela did not understand mathematical structures as uncreated entities that just exist. He would not concede that Stephen Hawking's question "What is it that breathes fire into the equations and makes a universe for them to describe?" had any real point at all. Unlike Tegmark who approvingly quoted Hawking on this issue, he would have denied that this is even a valid question. For Doniela, the basic structure of the universe consists of objects locatable within space and time. Physics and the laws of physics are descriptive of the way in which this universe behaves. The same applies to logic and mathematics. The laws of these latter disciplines, just as much as those of the former, explain how the universe in fact behaves or, to speak more accurately, how things within the universe actually relate to one another. None of these laws exists apart from the things described, except in the weak sense of having been conceptually abstracted from the objective states of affairs in which the laws alone are manifested. In short, if there were no universe, there would be neither physics, nor logic nor mathematics.

Contra to Hawking and contra to Tegmark, mathematical structures—equations—do not exist in some sort of Platonic realm waiting to be instantiated in the physical universe. Such a realm of mathematical, logical and other universal entities is superfluous to Doniela's ontology. Equations, according to his ontological position, are nothing more or less than formalized descriptions of the ways in which the universe itself is structured. What is ultimately real for Doniela is the range of the very things that Tegmark dismissed as baggage. Doniela would surely have taken issue with Tegmark's use of the phrase "human concept" to describe things like atoms or stars. Surely these are things and not concepts even though we form concepts of them. These entities existing out there in space and time relate equally to humans, super-computers, aliens and whatever else there may be and it is in virtue of their objective nature that the logical and mathematical structures in which they relate to one another must also be understood as existing objectively and independent of the human mind.

In short, although there is little doubt that Doniela would be happy to affirm that we live in a mathematical universe (he might wish to stress that it is also a logical one), there is considerable doubt that he would want to say that the universe is mathematics, unless by that expression he meant something a lot less radical than Tegmark implied.

Interesting questions arise from Doniela's philosophical position. For example, if space had a radically different geometrical or topological form, how much difference would that make to the laws of logic and mathematics? Do the laws of logic and mathematics continue to hold at distances smaller than the Planck length (10^{-33} cm) where space and time as we know them break down into quantum foam? If, we may speculate, space and time break down at this scale, both mathematical and logical structures must likewise lapse. But if that is true, we have the really weird situation where logic and mathematics lapse for the universe itself immediately following the Big Bang, at a time when its radius fell within the Planck length. And if logic and mathematics break down at scales approximating the radius of the universe immediately following the Big Bang, it presumably follows that the laws of physics also lapse at these scales. But does that place the beginning of the universe beyond the scope of physics?

These questions raise fundamental issues of physics, philosophy and indeed theology, discussion of which would take us beyond the scope of this book (although related issues have already been touched upon in *Weird Universe*). But the interesting point to note here is that they can legitimately be asked if mathematical structures are ontologically secondary to the material universe, though not if they exist uncreated and unchanging in the manner of Platonic forms.

If Doniela is correct and mathematics is descriptive of certain types of structure within the actual universe, there is no compulsion to postulate that we live in a level IV multiverse. Although a multiverse of this nature is possible according to his position, it is not necessarily implied by it. Unless there is sound evidence for the existence of this form of multiverse, which we must admit has not been forthcoming to date, its existence remains entirely within the realm of speculation.

Nevertheless, there is a sense in which Doniela's conception of mathematics and logic leaves open the theoretical possibility of something weirder even than a level IV multiverse. Earlier, we mentioned that, if logic and mathematics is ontologically posterior to the universe, neither would exist without the existence of the universe itself. Following this line of thought, we might say that, had the universe been somehow different, mathematical and logical structures within it need not have been the same as those with which we are familiar. But if there truly is a multiverse, that at least opens the theoretical possibility of the existence of other universes that may differ from ours in this manner and may indeed possess alien logics and mathematics. We cannot even imagine what such a universe would be like, simply because our minds are constrained in their thought by the logical and mathematical structures of our universe—which is why logic and mathematics came to be described as laws of thought in so many books on the subject and why the spurious problem as to how they manage to describe the universe so accurately arose. Doniela's thesis does not imply that such alternative universes exist. It only allows them to be, as we said, theoretically possible.

As a final word on Tegmark's hypothesis, we may say that although we find it unconvincing, it at the very least puts mathematics in its correct place (the external world) and avoids the creeping subjectivism that so often haunts speculation concerning the connection between mathematical equations and the material universe. Moreover, any hypothesis that forces us to think outside of the proverbial box and that stretches our intuitive responses to new ideas can only be beneficial and, even if it ultimately proves to be incorrect, may still direct our thinking in the direction of a more accurate theory of reality. For these reasons alone, Tegmark's mathematical universe hypothesis constitutes an important contribution to both scientific and philosophical thought.

Of Strings and Other Things

As every school student learns, like electrical charges repel and unlike ones attract (Fred Hoyle's suggested modification earlier in this chapter notwithstanding!). But this simple fact presented atomic physicists with a problem. Except for the simple atom of hydrogen with its single proton, all atomic nuclei possess more than one positively charged particle. Why, therefore, do they not fly apart as the (like) electrical charges of the protons repel one another? Why, in other worlds, is matter stable? We would think that the only stable atoms would be those of hydrogen. If electricity had the final word therefore, the universe would be a very dull place indeed; just one vast cloud of hydrogen without stars, planets and life!

Clearly something other than the electrical force (or, as we should more correctly call it, the electromagnetic force) is present within the nuclei of atoms and this force, or whatever we might like to call it, is a good deal stronger than electromagnetism, at least over distances comparable with that of an atomic nucleus.

Back in 1968, CERN physicist G. Veneziano began working on a theory of the atomic nucleus which, together with further refinements and developments by L. Susskind, H. Nielsen and Yoichiro Nambu during the following 3 years, initially seemed to offer the long awaited solution to this problem. What these physicists proposed was the existence of minuscule vibrating filaments of energy having lengths of just one Planck Length or 10^{-33} cm and tensions equal to one Planck Force (10^{44} N); tiny "strings" of energy so to speak, which in effect tied together the protons and neutrons within an atomic nucleus. The strings were strictly linear or one-dimensional; they literally had no "thickness" whatsoever. The nucleons (that is to say, the positive protons and the neutral neutrons) could be thought of as attached to the strings' endpoints and in this way tethered together. Of course, this picture is an extremely crude way of thinking about the actual situation, but as we can only experience, and thereby form mental images of, the world at macroscopic levels—macroscopic at least in comparison with the sub-atomic realm—and as the latter is qualitatively as well as quantitatively different from the realm of familiar experience, any attempt to imagine what the process within an atom looks like must be symbolic at best. Needless to say, the string theory of the atomic nucleus as derived by Veneziano and his colleagues was presented in mathematical and not pictorial terms, but the crude picture given here at least can give some vague notion of what it entailed (Fig. 5.5).

Unfortunately, although it initially seemed to be satisfactory in a general sort of way, it was not long before problems began to appear in the finer details. Predictions made on the basis of the theory did not accord with experimental evidence and by the early 1970s it was becoming clear that an alternative theory was required.

Actually, the foundations of this alternative theory had already been laid back in 1964 in the form of a model put forward by physicists Murray Gell-Mann and, independently, George Zweig. According to these scientists, nucleons are not fundamental particles, as previously thought, but are instead composed of

Fig. 5.5 Professor Gabriele Veneziano (*Courtesy*: Betsy Devine)

point-like particles to which the name "quark" was given. This theory did not immediately take the world of physics by storm. Nevertheless, a few physicists continued to work on the model during the 1960s. In particular, the theoretical developments made by Moo-Young Han and Y. Nambu during that decade demonstrated how quarks could be held together by a class of massless particles which came to be known as "gluons" and which were thought of as the carriers of a strong fundamental force that acted over very short distances (Fig. 5.6).

All of this stayed within the realm of pure theory until 1968 when the Stanford Linear Accelerator supplied the first experimental evidence that quarks really do exist. We might recall that, coincidentally, this was also the year that Veneziano began to develop the string model of the atomic nucleus.

In 1971, Gell-Mann and H. Fritzsch proposed that the Han/Nambu model should be taken seriously as an alternative to the string theory. Their theory, as already mentioned, depended upon the existence of a fundamental force manifesting powerfully at distances comparable with the radii of sub-atomic particles, yet fading to insignificance beyond the scope of an atomic nucleus. Three

FIG. 5.6 Professor Murray Gell-Mann (*Courtesy*: Charlotte Webb)

varieties of quarks were proposed, differing from one another in a property which bore some resemblance to electromagnetic polarity in the sense that likes repelled and unlikes attracted; except that the force was far stronger than electromagnetism and there were three charges instead of just two. Although this force is totally unlike anything that we experience in the macroscopic world, the fact that it came in three types suggested an analogy of sorts with the three broad kinds of colors perceived by humans, that is to say, red, green and blue. Therefore, for that reason alone, the three quark types, charges, or whatever else we might like to call them, were given the term colors and the force between them became known as the color force. Briefly, quarks that have different color charges attract and those of the same color repel.

This color force has two manifestations. At the strongest level and over distances smaller than that of the radius of a proton it binds quarks together into nucleons. In this manifestation it is also known as the strong interaction and is carried by the gluons which, as already mentioned, are without mass and therefore travel at the speed of light.

The second and larger scale on which the color force is manifested is the one which brought the issue to light in the first place, namely, its manifestation as the binding together of nucleons within the nucleus of an atom. Although known as the strong force this is in reality a weaker manifestation of the color force. It is actually a residual force, albeit still powerful enough to cancel out the electromagnetic repulsion of protons, i.e. those nucleons with the same positive electric charge. Curiously however, at the very smallest distances, less than about 0.6 proton radii, the force becomes repulsive and prevents the nucleons from coming too close to one another. The larger range of the color force—the strong force—extends from a little over one proton radius to nearly four proton radii, after which it falls off into insignificance.

As developed in the 1970s, this theory accounted for the properties of atomic nuclei and was well supported by experimental evidence, free from the problems that beset its string-theory alternative. Because of the color terminology, the theory was given the moniker of quantum chromodynamics and from early in the '70s decade, replaced the string model as the accepted theory.

Nevertheless, although the string model of the strong nuclear force got itself into too many tangles and had to be discarded, that was not the last that physics would hear about strings. Indeed, the following decade was to see the theory rise, to use Shelley's words, "like a ghost from the tomb" to assume a position as one of the most prominent, as well as one of the weirdest, of physical theories.

Ironically, one of the biggest problems raised by the theory in its early application turned out to be the very cause of its recovery. As an attempted answer to the problem of the stability of atomic nuclei, one of the difficulties encountered involved the prediction of an unusual sort of particle which was not observed and apparently does not exist. Yet, in a different context, a particle with similar properties could prove to be very useful because, ironically, the troublesome phantom particle happened to have just the right credentials to play the role of the hypothesized graviton; the particle required for the elusive quantum theory of gravity. String theory may have failed to provide a satisfactory model of the atomic nucleus, but is it possible that it might yet become the foundation of a theory of quantum gravity? If that proves correct, the exciting prospect of a Theory of Everything—the theory that unites all of the forces of nature into a single model—might finally be realized.

According to string theory as it developed in the 1980s, strings can oscillate in a variety of ways. On scales greater than that of the string radius, each of these oscillation modes gives rise to a different species of sub-atomic particle, the mass, charge and general properties of which are determined by the dynamics of the string. Particle emission and absorption correspond with splitting and recombinations of the strings. Strings may be either open (these can be pictured as having open ends) or closed like a loop. These two types of strings behave in slightly different ways, but because the two ends of an open string can always join up and turn the open string into a closed one, all string models necessarily contain closed strings. Interestingly, all of these contain graviton modes of oscillation but only open strings are capable of oscillations corresponding to photons.

These oscillations—subatomic particles as we have traditionally termed them—may be analogously compared with the notes of a plucked guitar string. They can also be thought of as waves on the surface of the worldsheet of a string.

What is meant by a worldsheet? If we graph a point moving through space-time, we end up with a line known as a worldline. In effect, it is a graph of the spatiotemporal history of that point (refer above to Tegmark's bird perspective). If a similar graph is made of a string, the fact that the string is itself a one-dimensional line means that the resultant path is broader than a line. This is what is known as a worldsheet. Worldsheets of closed strings may be represented by the form of a pipe while those of open strings may be represented by the form of a strip.

In its earliest formulation, the revived string theory applied only to particles known as bosons, that is to say, particles similar to photons which roughly speaking constitute radiation rather than matter. This early model encountered difficulties in that it turned out to be fundamentally unstable. Moreover, by its very nature it failed to account for the existence of particles of ordinary matter. Further development and refinement widened the scope of string theory into a more inclusive model which included both bosons and fermions, i.e., the particles comprising familiar matter. In the process, a mathematical relation between bosons and fermions known as supersymmetry was discovered and the versions of string theory including both types of particles became known as supersymmetrical string theories or, more simply, superstring theories. Supersymmetry requires that for every particle, there must be a superpartner whose spin differs by one half that of the original particle. Although in one sense this might be said to complicate the particle scene, in a more important sense it promises to simplify it by holding out the promise of a theory that encompasses both bosons and fermions.

As it developed in the latter years of last century, superstring theory became the hope of many physicists in search of the thus far elusive Theory of Everything. It began to look like a promising candidate, as it not only provided something that looked tantalizingly like a graviton, but also seemed to account for all the particles of the standard model of particle physics as well as the forces which operate upon them.

A potential difficulty—in trying to picture the strings and their oscillations, if not necessarily in the theory itself—is that these string movements would need to take place in a hyperspace of many dimensions to account for the required phenomena. The early bosonic (bosons only) version required as many as 26 space-time dimensions. Fortunately, the more promising superstring theories only needed 10. Nevertheless, if we really are living in a universe with ten dimensions, the big questions desperately needing answers are why we only experience three dimensions of space and one of time, and where are the rest hiding.

One possible answer to this question lies in the suggestion that these extra dimensions—by contrast with those defining our familiar macroscopic realm—are compactified to very small values. To take a familiar analogy, a length of very thin cotton thread looks almost like a two-dimensional line, although we know that it is in fact a three dimensional object. It merely seems to be two-dimensional because it is extended for only a very short distance in the dimension of width as compared with that of its length.

A second possibility is that we are trapped in a subspace of the universe that manifests just three spatial and one temporal dimension. Just as astronomers have discovered that the matter with which we are familiar constitutes only a small fraction of the universes' total mass, so this theory proposes that the dimensional space with which we are familiar constitutes but .a minor portion of the whole. It is not impossible that spacetime contains strata of various dimensions and that we inhabit a stratum consisting of three spatial/one temporal dimensions that may not be typical of the whole.

There is also a third possibility which, although suggested by no less a physicist than Steven Weinberg has nevertheless not been widely canvassed. This is to, in effect, "de-literalize" the dimensions by interpreting them more as mathematical fictions than as actual properties of the real universe. As Sten Oldenwald expressed it in his book *Patterns in the Void: Why Nothing is Important*, "These extra dimensions may not have direct physical meaning at all, at least not like space and time do. There are three flavors of space (up, down, and sideways) and one flavor of time. Strings require six more flavors of something else. The color of a quark, in the special sense of color used in quantum chromodynamics, is in some ways a dimension of the particle, but certainly not one you can measure in inches or miles." (p. 182). The hint here is that the extra dimensions required by string theory may no more resemble either the spatial dimensions or the temporal dimension of the macroscopic world than the quark colors, which form the subject of quantum chromodynamics, resemble the color of familiar trees and flowers.

A more serious problem however is that string theory developed into five different varieties. That was not a good sign for something that was supposed to explain all of nature within the parameters of a single theory.

Equally distressing for those hoping for a unique theory of everything is the embarrassing fact that string theory predicts the existence of some 10^{500} possible alternatives from which our universe was, in a manner of speaking, picked. This number is so stupendous that it in effect means that just about any phenomenon which might be observed at lower energies could be described by some form of string theory, rendering the model untestable as it stands.

The Branes of the Universe

By the middle of the 1990s, superstring theory had been stuck in a mathematical bog for several years and seemed in danger of sinking out of sight. With five rival theories each having as much and as little claim to being the correct one, the situation had clearly arrived at a stalemate. Then in 1995, a breakthrough came when the brilliant physicist Edward Witten demonstrated that these five apparently disparate theories were really just various aspects of an even more fundamental one, now known as M-theory. In Oldenwald's words "The resolution resembled what would happen if a group of physicists wandered around in a chicken-packing plant and found a leg here, a thigh and a breast over there. Each would seem like a unique object. But M-theory put the pieces together into a whole chicken." He then optimistically added "If we listen carefully, this particular chicken may tell us the secrets of the universe." (p. 183). This last comment reflects the hope of many within the physics community that M-theory, when developed more fully, will either be, or become a significant step toward, that great prize of theoretical physics, the Theory of Everything.

Central to M-theory is the concept of a brane. The word is derived from membrane (hence the M) and is used to characterize an object of a certain number of dimensions when considered from a higher dimensional viewpoint. It is a subspace of a larger total space, the latter known as the bulk. For instance, a point particle is a 0-brane, a string is a 1-brane and something like the surface of the ocean might be termed a 2-brane wrapping the Earth and propagating in the four-dimensional spacetime of the Solar System environment. The screen of a computer might likewise be termed a 2-brane existing within the bulk of our familiar three-dimensional space.

By introducing this concept into string theory, Witten is taking the second of the abovementioned alternatives around the extra dimensions problem. We do not experience the full number of dimensions because all material particles and forces are restricted to a three-dimensional space called the 3-brane. In effect, our universe lies on (or we could equally say "is trapped within") the three-dimensional "surface" (something like a membrane) of a bulk space possessing a larger number of dimensions. Indeed, M-theory requires one more dimension than superstring theory had earlier demanded, but if a ten-dimensional universe is accepted, adding an eleventh hardly raises any difficult issues, especially if it really does hold out the promise of a truly comprehensive theory! For Witten, these extra dimensions are just as real as those of our familiar world. He does not appear to think of them in the non-literal sense proposed by Weinberg. On the contrary, the real universe is seen as an 11-dinensional bulk in which float islands and continents of fewer extended dimensions. Our universe is one of the 3-D islands, but there may be many others as well; some with the same number of dimensions as ours although we also cannot rule out other islands or branes that possess more (or maybe fewer?) dimensions and that may even be extended macroscopically in some of the dimensions that are inaccessible in our universe. In fact, even our universe may have a limited extension into some of these other dimensions, according to Princeton University's Lisa Randall and Raman Sundrum. Dimensions as large as the Solar System or as small as a speck of dust could be associated with out 3-brane without having a noticeable effect on the ones with which we are all familiar.

The possibility that other 3-brane universes exist raises the weird prospect of universes similar to the one we know existing very close to our own; just separated from ours by a tiny distance along some dimension of which we can have no experience. For all we know, there may be a world not unlike planet Earth located, in one sense, almost at the same place as our own while in another sense infinitely remote. Because everything we see in our universe is confined to the one 3-brane, no matter or radiation (with one possible exception that we shall look at in a moment) can pass from our universe into a universe confined to a neighboring 3-brane, no matter how close that neighbor is as measured along the separating dimension. Assuming that their matter and radiation is similarly confined to their 3-brane, no matter or radiation can pass from them across to us either.

An interesting suggestion has been put forward in connection with some versions of M-theory, namely, that whereas, as we stated above, all the strings that represent the particles of matter and radiation begin and end on our 3-brane, maybe gravitons alone have a greater freedom. In fact, they may be free to roam the entire 11-D bulk. Gravity, in other words, may actually leak out of our 3-brane and into the wider universe and this leakage may be the reason why gravity is such an exceptionally weak force in comparison with the other fundamental forces of nature.

Although watching apples fall from trees and planets orbiting the Sun may disincline us to think of gravity as a weak force, the fact remains that when it is compared to the electromagnetic and nuclear forces, it is extremely weak indeed. Suppose that we designate the strength of the strong atomic force as 1, the electromagnetic force would then register as 1/137, the weak atomic force as 10^{-6} but gravity would come in at an incredible feeble 10^{-39}. But why should gravity be so weak? Physicists have long been puzzled by this question. While a much stronger gravity would preclude the existence of living organisms of any kind, it remains puzzling.

If however gravity exists in multiple dimensions, it need not be intrinsically so weak after all. Extension into other dimensions will also alter the way in which gravity works, however if it really is free to roam across all of the dimensions, the scale on which it will differ from what is predicted for a strictly three-dimensional gravity will be too small to observe. For example, if we suppose that gravity leaks into just one further dimension, the scale on which it would differ from "standard" three-dimensional gravity would be apparent over the sort of distances covered by solar-system scales. For a leakage into two further dimensions however, the scale shrinks to around that relevant to fleas and for leakage into three extra dimensions, the scale shrinks to subnuclear size.

If gravity really does cross the dimensional barriers, it may truly be a means of communication between our 3-brane (sub)universe and other neighboring (sub)universes. Violent events in our 3-brane—black hole collisions or stellar-core-collapse supernova for instance—should radiate a burst of gravitational energy into the multi-dimensional void. This might in theory be detected in neighboring 3-brane universes. Events in a neighboring 3-brane, or even a brane of more dimensions, might be detectable in our universe and vice versa. One might even dare to speculate that highly advanced civilizations in different 3-branes might find a way of contacting one another by gravity-wave radio or the like.

Coming back to Earth, or at least to our 3-brane, note that if some of the gravitational energy in a violent cosmic event such as a core-collapse supernova leaks out of our brane, this should be detectable to astronomers as a discrepancy in the energy balance of that supernova explosion. To date, no such missing energy has been suspected in supernova explosions; not even in the well observed naked-eye object 1987A in the Large Magellanic Cloud.

These speculations about gravity are fascinating, but lack evidence at this time. In fact, the apparent lack of missing energy in supernova explosions does not look promising for the hypothesis. Nevertheless, even if it turns out that gravity is no freer than any of the other forces of nature, this will be no serious detriment to M-theory per se. This model of gravity requires M-theory, in some form or other, to be correct but M-theory per se does not necessarily predict this multi-dimensional characteristic of gravity.

The same may be said for another interesting hypothesis based upon the model of the universe implied by M-theory. This is the cosmological model proposed by Neil Turok and Paul Steinhardt which postulates a region of five-dimensional space-time in which the four spatial dimensions are bounded by two 3-branes, rather like two three-dimensional walls. One of these walls is the space in which we live. Between these two, there is a third 3-brane which, in effect, is loose and very occasionally hits one of the boundary 3-branes. A little less than 14 billion years ago, it crashed into our brane, releasing a great deal of heat through the effects of the collision. This burst of heat was none other than the Big Bang, although a somewhat different Big Bang from the one usually hypothesized. For one thing, it did not begin in a singularity or in some ultra-dense state approaching a singularity. It also did not require the extremely brief burst of exponential spatial expansion known as inflation that has for quite a number of years been an integral part of orthodox cosmological models.

The Turok/Steinhardt model has officially become known as the ekpyrotic universe theory or, more colloquially, as the Big Splat. It is considered a cyclic model as the loose brane is thought to periodically impact the 3-brane that currently houses our universe, although the interval between collisions is long by comparison with the age of the visible universe itself.

There are some attractive features of this model, principally its avoidance of that troublesome initial singularity, however its inconsistency with the existence of an inflationary era presents a problem. Even though several physicists argue that inflation has been made to carry loads that it cannot properly bear, the theory itself still has a lot going for it.

During 2014, a possible astronomical discovery was announced that, if confirmed, would have all but clinched the case for inflation and sunk that of the ekpyrotic universe. Astronomers have for several years believed that if there is a smoking gun of inflation it lies concealed in the Cosmic Microwave Background (CMB). Patterns in the CMB show clear evidence of structures that formed very early in the history of the universe and acted as templates for the chains of galaxies and large voids that populate the visible universe today. The discovery of these features was a remarkable advance in our understanding of the early cosmos, but what astronomers sought even more diligently was the pattern of a kind of wave that is predicted to have arisen during the inflationary era. These are the gravitational waves predicted by general relativity, effectively waves in the fabric of the space-time continuum itself. They are believed to be produced in abundance by violent events such as the collapse of a star into a black hole, but they are very difficult to detect and have not actually been observed directly although the strength of relativity theory and some indirect astronomical evidence for their presence leave little room for any serious doubt about their presence in the real universe.

Gravitational waves should also have rippled through the early universe following the Big Bang and it is theoretically possible to detect these through the polarization of the CMB. There are, however, two modes of polarization of the CMB. One, which has been known for several years, is the E-mode polarization and does not arise from gravitational waves. The type produced by these is known as B-mode, but there is a slight complication insofar as E-modes can be converted to B-modes through the gravitational lensing of galaxy clusters interspersed between Earth and the CMB. Therefore, even without inflation, a portion of the E-modes will be picked up on Earth as B-modes. Fortunately however, this secondary B-mode polarization can be distinguished from what we might call the primordial B-modes arising directly from gravitational waves.

The first variety of B-mode polarization was discovered by the South Polar Telescope team and officially announced in July 2013. This was an important discovery of course, but it said nothing directly about the existence of the inflationary era. Nevertheless, continued observation at the same observatory led to the big announcement, in March 2014, of the first detection of primordial B-modes in the CMB sky. Needless to say, this announcement was greeted with a wave of excitement sweeping through the cosmological community. Had the smoking gun of inflation finally been found?

Alas, that does not appear to have been the situation. Further investigation suggests that this was a mistaken observation and the apparent B-modes detected resulted from the relatively local source of dust within the Milky Way galaxy. The issue is still being examined in case real primordial B-modes are buried within the noise, but it does appear that the claim was premature and that inflation's smoking gun has not after all been detected.

It should be stressed that the ekpyrotic model, although obviously dependent upon the validity of M-theory, stands apart from it in the sense that it is a model postulated within the M-theory framework and not a necessary consequence of that theory per se.

The universe could be described accurately along M-theory lines and yet not have the floating brane hypothesized by Turok and Steinhardt. Or is it necessary that the floating brane, if it really does exist, should repeatedly bounce off another brane? Earlier models of a cyclic universe (an eternal series of big bangs and big crunches) ran into trouble in so far as there was a loss of energy at each bounce. Unless the hypothesized brane collisions are perfectly elastic, would that not also apply to them as well?

It is now around 20 years since M-theory was first put forward and in that time it has received the attention of some of the most brilliant mathematical minds on the planet. Yet, it is still in what might best be described as a preliminary phase. The mathematics is extremely difficult and there are many issues that are yet to be worked out. Moreover, like superstring theory in general, it is not easy to verify or falsify through observation and/or experiment and this fact alone does not go down well with certain physicists. Maybe the difficulty of the theory is a stroke against it; a symptom that it is a red herring drawn across the path to a true Theory of Everything. Yet, on the other hand, while we may well agree with Einstein when he said that the strangest thing about the universe is that it is comprehensible to human beings, we must also acknowledge that neither Einstein nor anyone else declared that this comprehensibility would come easily. As the ancient Greek proverb says "Truth lies at the bottom of a deep well" and we might expect that the more profound the truth, the deeper will be that well and the narrower the shaft such that only those proficient at the most difficult math will be able to slide down it. M-theory appears to fulfill this expectation, so perhaps its very difficulty is a sign of its essential validity.

M-theory stands upon the shoulders of superstring theory which, in its turn, stands upon the shoulders of supersymmetry. Of these three levels, the last mentioned is the one most amenable to verification/falsification with today's technology. If this could be verified, M-theory would certainly be strengthened, although not necessarily proven. Nevertheless, even if it should turn out that supersymmetry is not correct, the whole pyramid would not necessarily come crashing down, as alternative versions of string theory are possible and it may be that an alternative to supersymmetry would open up new ways in which string theory and M-theory could be tested. Still, verification of supersymmetry would be seen by most contemporary workers in the field as an indication that they are probably on the right path. The discovery of superparticles—the ultimate verification of supersymmetry—should be within the capability of the Large Hadron Collider and it has been the hope of the physics community that this wonder of science and technology would provide the evidence so long sought. The situation has parallels with the Higgs boson, which was eventually discovered by the LHC. As for superparticles however, at the time these words are being written in the middle of May 2015, the hope remains unfulfilled. Maybe this will change with the next run of the LHC. Maybe the discovery will be made by the time you read these words. Or maybe not! At the moment though, physicists working on the supersymmetry issue are becoming uneasy although not yet desperate.

One positive LHC result has probably been as responsible for this unease as the negative ones, the failures to find superparticles, have been. In 2012 and 2013, the LHC witnessed the extremely rare decay of the Bs meson into two muons at a rate that was found to be exactly as predicted by the Standard Model of particle physics. Now, normally when an experimental or observational result is in good agreement with the prediction of a theoretical model, scientists rejoice. But not this time. The problem is that nobody really believes that the rather ad hoc Standard Model has the last word in particle physics. Everyone agrees that there must be some physics beyond it, and the most popular choice for this new physics is supersymmetry. Had the LCH results differed somewhat from those predicted by the Standard Model, this would have been the first clue that new physics beyond the model does indeed exist and would have provided at least some indirect support for supersymmetry. That this failed to happen was seen by some as a blow (though hopefully not a fatal one) to supersymmetry. A news note in May 2015 indicates, however, that the decay of another form of B meson, namely neutral B mesons, into two muons was observed to occur at about four times the rate predicted by the Standard Model. If this is confirmed, it does indeed hint at a physics beyond that of the Standard Model, although this may not necessarily equate with supersymmetry and in fact doubt has already been raised that it does. Something even weirder might be implied by these new results.

This, then, is the situation at present. Whether M-theory or some near approximation of it turns out to be the long awaited Theory of Everything or a further step in the direction toward such a theory, or whether it is finally revealed as nothing more than a distraction from where the real theory lies, it will certainly be long remembered as a weirdly brilliant cosmological model.

Appendix A: The Maribo Meteorite

Although the Maribo meteorite may ultimately have come from Comet Encke, or the hypothetical comet from which Encke and the Taurid meteor complex is widely thought to have originated, the more recent history of this object appears to associate it with several other minor Solar System bodies. As mentioned in the main text, there appears to be a strong association (as determined by Drummond's D' criterion) with the Apollo asteroid 85182. But the associations do not cease there.

In their paper published in *The Observatory* (1994 October), Duncan I. Steel and David Asher distinguished two groups of Apollo asteroids; the larger of the two pursuing orbits similar to those of the Taurid meteors and the smaller having orbital elements similar to those of the asteroid (2212) Hephaistos. These apparent families were simply called the Taurid group and the Hephaistos group respectively. As noted in Chap. 1 of the present book, each group also includes a comet; Encke in association with the former and Helfenzrieder in association with the latter. Encke remains an active object, but Helfenzrieder was observed (as a relatively bright naked-eye object sporting a long tail) only during its 1766 apparition. This object was probably dormant at earlier returns and possibly broke apart in 1766, going out in an uncharacteristic blaze of glory. Any remnant that may continue to exist today is presumably small and very faint although some day it may be discovered anew by one of the search programs seeking near-Earth objects.

Steel and Asher found that the two asteroid groups distinguished themselves most readily through the respective ranges of longitude of perihelion of their members' orbits. This value is the sum of the argument of perihelion and longitude of the ascending node of the orbit and can be thought of as the orbit's orientation. Taurid group orbits have longitude of perihelion values between 100° and 190° whereas Hephaistos group orbits fall between 222°

© Springer International Publishing Switzerland 2016
D. Seargent, *Weird Astronomical Theories of the Solar System and Beyond*, Astronomers' Universe, DOI 10.1007/978-3-319-25295-7

and 251°. These limits have no magical significance: they are simply the values between which the objects of each group that were known in 1994 fell. Steel and Asher opined that other objects falling within the gap may eventually be found, effectively turning the two groups into concentrations at each end of a single extended family.

This latter scenario would actually fit with what Steel and Asher see as the dynamical history of these groups. With respect to perihelion distance, eccentricity and inclination, the orbits of the members of both groups are at the same time similar to one another and yet different from the typical Apollo asteroid. As argued in the main text of the present book, there are reasons to think that a number of the objects included within the Taurid group might, however, by interlopers, but that does not alter the basic model. Some, at least, of the suspected interlopers have orbits which are less typical of the Taurid group than those of the "core" members. In any case, the atypical orbits of members of both groups are best explained, according to Steel and Asher, if the entire complex originated in the prehistoric disruption of a large periodic comet. The comet initially split into two major pieces, each of which progressively broke up over many perihelion passages. The Taurid group represents the fragments of the principal nucleus and the Hephaistos group consists of the fragments of the main secondary comet. Presumably all of the asteroids were once active comets that have now become dormant through the build-up of an insulating layer on their surfaces. Maybe Encke was dormant for many years but became active again prior to its discovery through (we might suppose) the fracturing of its insulating layer. Helfenzrieder may have suffered a worse catastrophe, activating in an extreme way at a single return and then fading out completely.

As mentioned in Chap. 1, the Maribo meteorite appears to be a member of the Hephaistos group. Its longitude of perihelion is 217°, a little toward the Taurid side of the lower limit given in the Steel/Asher paper although, as we said, there is nothing absolute about this lower limit. Their lower value appears to have been set by the asteroid 85182 (222°) which we suggested is the immediate parent body of the meteorite.

Further examination yielded an even bigger surprise however. This asteroid seems to be the parent object of the Delta Cancrid meteor shower and the Maribo meteorite is a member of this shower. If this is confirmed, it represents the clearest association yet discovered between a meteor shower and a meteorite.

Taking the orbit of the shower as being the average of three orbits published by Gary Kronk in his *Meteor Showers: A Descriptive Catalog* (1988), we find the following;

Longitude of perihelion of Delta Cancrids = 222.3°
Delta Cancrid/Maribo meteorite, $D' = 0.0717$
Delta Cancrid/asteroid 85182, $D' = 0.0821$
Maribo/asteroid 85182, $D' = 0.04$

Looking at the three individual orbits noted by Kronk;

S1973/asteroid 85182, $D' = 0.0803$
S1976/asteroid 85182, $D' = 0.115$
L1971B/asteroid 85182, $D' = 0.0595$
S/1973/Maribo, $D' = 0.066$
S1976/Maribo, $D' = 0.1086$
L1971B/Maribo, $D' = 0.0379$

The orbit for the southern Delta Cancrids noted by Kronk did not fare so well, yielding D' values for 85182 and Maribo of 0.1651 and 0.1581 respectively. Considering the degree of scatter expected in this meteor stream however, that is probably not too surprising.

It is interesting to note that Maribo fell on January 17, the date usually given for maximum of the Delta Cancrid shower. Moreover, the velocity of the Delta Cancrid meteors has been measured as 28 km/s in excellent agreement with that of Maribo.

Orbits having Drummond D' values of 0.105 or less are indicative of association. The case for the above suggested associations, therefore, appears to be rather strong.

Appendix B: The Sutter's Mill Meteorite: A Taurid Connection?

In terms of its longitude of perihelion, high eccentricity, small inclination and a perihelion distance near the orbit of Mercury, the orbit of the Sutter's Mill meteorite is indicative of a Taurid association. Yet, that is not obvious from its date of fall; April 22, 2012. A search of meteor showers published by Kronk nevertheless uncovered something interesting. He lists a daytime shower (May Arietids) extending from early May until early June and gives three orbits as determined by radio observations made during the 1960s. All of these orbits show some similarity with that of Sutter's Mill. The early date of the meteorite does not necessarily count against association with the stream, as it is not unusual for weak activity to extend beyond the limits normally given for meteor showers. Taking an average of the orbits given by Kronk and comparing this against that of Sutter's Mill yields a D' value of 0.097. Equally intriguing is the comparison between the average May Arietid orbit and the one computed by Kronk for the southern Taurids. This comparison yields D'=0.094.

The May Arietids probably have a southern and a northern branch and the orbits given by Kronk seem more representative of the northern branch. From the apparent association with the southern Taurids, it appears that this northern branch of the May Arietids (though not as well established as the southern one) forms part of the same stream as the southern Taurids. When Earth encounters the stream members on their way toward perihelion—which happens during October and November—those meteoroids which then enter our atmosphere are observed as southern Taurids. When Earth intercepts part of the same stream on the outward journey around May and June, some encounter our planet as northern May Arietids. In this connection, it is interesting to note that Sutter's Mill arrived at perihelion just over six weeks

© Springer International Publishing Switzerland 2016
D. Seargent, *Weird Astronomical Theories of the Solar System and Beyond*, Astronomers' Universe, DOI 10.1007/978-3-319-25295-7

prior to its encounter with Earth, consistent with having been an early-arriving member of this shower.

The immediate parent of the southern Taurids is considered to be Comet Encke. Presumably, this comet is also the immediate parent of, at least, the more northerly members of the May Arietids. If that is correct, it seems that a good case can be made for identifying this meteorite as also being a fragment of Comet Encke.

Author Index

© Springer International Publishing Switzerland 2016
D. Seargent, *Weird Astronomical Theories of the Solar System and Beyond*, Astronomers' Universe, DOI 10.1007/978-3-319-25295-7

Subject Index

A

Ambiplasma, 64, 65
Antimatter, 65, 83, 169–174, 188
Apollo asteroid, 31, 67, 68
Asteroids
   1979 VA (*see* Comets,
     Wilson-Harrington)
   1989 VB, 38, 44, 46
   1991 AQ (=1994 RD = 85182), 30
   1996 RG3, 182
   1997 YM3, 44
   2000 PF5, 44
   2001 PE1, 44
   2008 TC_3, 115
   active, 6, 16, 31, 32, 37, 38, 129, 131,
     135, 136, 143, 150, 152, 160, 164,
     166, 182, 183, 185, 197, 200
   Anza, 38, 44
   Baptistina, 114–117, 124
   Ceres, 74
   Flora, 116
   Hephaistos, 30, 31
   Nemesis, 124
   Pallas, 166
   Phaethon, 151, 166
   short period, 130

B

Bacillus infernus, 5, 91
Bacteria, 13, 26, 48, 50
Bacteriophages, 50
Beta Taurid meteor, 185
Big Bang cosmology, 4, 65, 66, 204, 205,
   208, 209, 225, 239, 252–254
Black holes, 157, 169, 227
Branes, 248–256
Buddhism, 196, 197, 200

C

Canterbury, 188, 189, 191, 192
Carbonaceous chondrites, 33, 38, 42,
   113, 115, 116
Cataclysmic variables, 80
Catastrophism, 61
Centaurs
   Chiron, 79
   Echeclus, 139, 140
C-gravitons, 84
Chicxulub crater, 107, 110
Chiron, 139, 140, 180, 181
Chiron's dimensions, 181
Chondrules, 168
Color force, 244
Comet
   214 BC, 154
   302AD, 154
   467, 154
   Atlas, 195, 196
   Bennett, 160, 197
   Biela, 36, 67
   Churyumov-Gerasimenko, 140
   d'Arrest, 152
   Denning-Fujikawa, 44
   Donati, 160, 161
   Encke, 29, 30, 45, 49–52, 150, 177,
     181–184
   Finlay, 30, 38–41, 43–47
   Hale-Bopp, 145, 180
   Halley, 45, 51, 146, 149, 162, 198
   Haneda-Campos, 44, 45
   Hartley, 140
   Helfenzrieder, 30
   Holmes, 152, 199
   ISON, 78
   Kohoutek, 161
   Lovejoy, 155, 156
   Machholz, 187
   Metcalf, 32, 33
   Pons-Winnecke, 32, 33
   Schwassmann-Wachmann, 199
   short period, 28, 30, 32, 34, 38, 41, 46,
     49, 128, 134–137, 140, 141, 143,
     144, 155, 166, 176, 187, 198, 199
   showers of, 31
   SOHO (May 1999), 186–187
   sungrazing, 153, 154
   Swift-Tuttle, 69
   Tempel-Tuttle, 69

© Springer International Publishing Switzerland 2016
D. Seargent, *Weird Astronomical Theories of the Solar System and Beyond*, Astronomers' Universe, DOI 10.1007/978-3-319-25295-7

Printed in the United States
By Bookmasters